AF348970

To: Pippa, Fletcher, and Vesper

Ethics of Innovation in Neurosurgery

Marike L. D. Broekman
Editor

Ethics of Innovation in Neurosurgery

 Springer

Editor
Marike L. D. Broekman
Department of Neurosurgery
Haaglanden Medical Center & Leiden
University Medical Center
The Hague and Leiden
Zuid-Holland
The Netherlands

ISBN 978-3-030-05501-1   ISBN 978-3-030-05502-8   (eBook)
https://doi.org/10.1007/978-3-030-05502-8

Library of Congress Control Number: 2019930129

This Springer imprint is published by the registered company Springer Nature Switzerland AG
The registered company address is: Gewerbestrasse 11, 6330 Cham, Switzerland

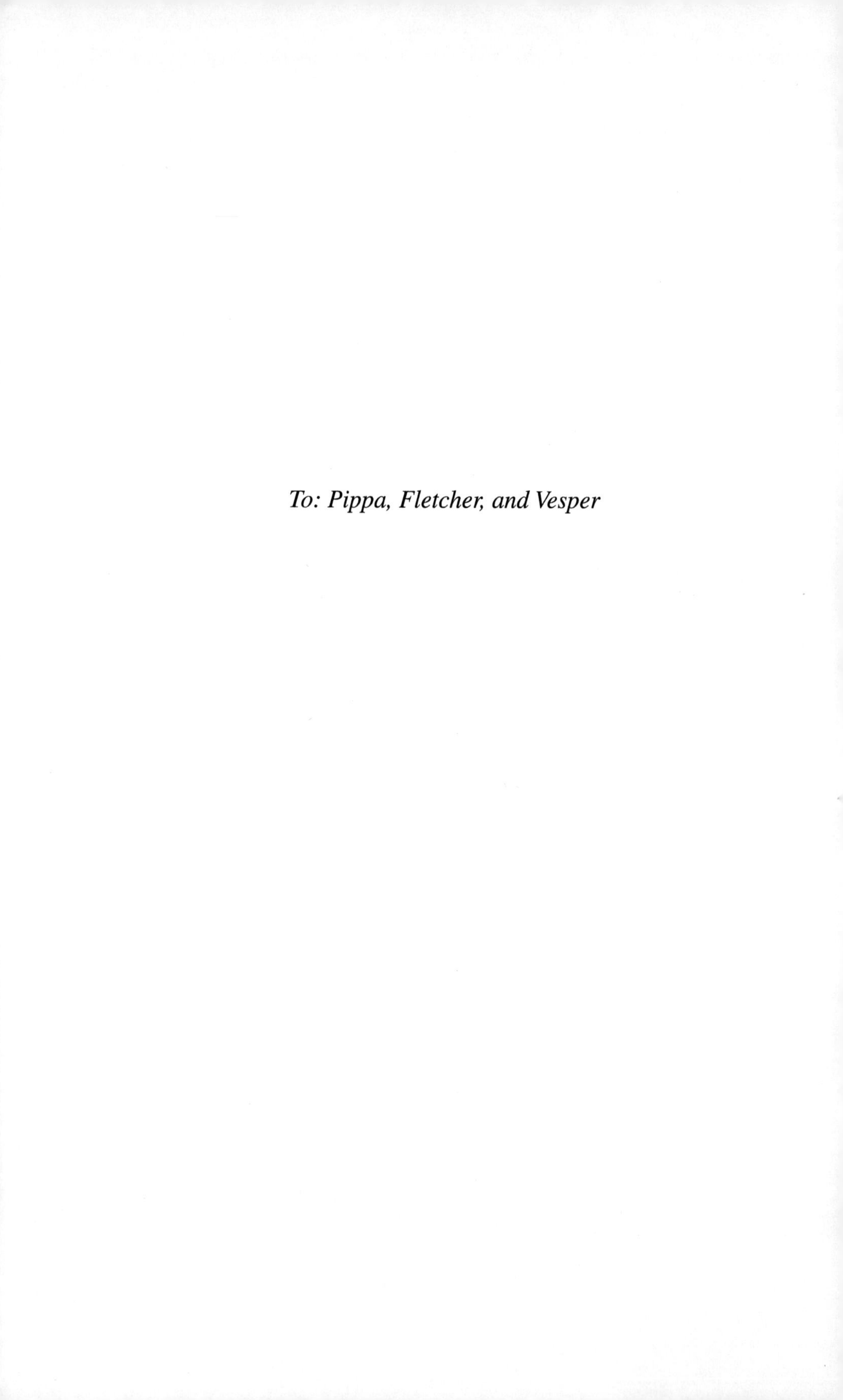

To: Pippa, Fletcher, and Vesper

Innovation is at the heart of neurosurgery. If great surgeons had not been given room for innovation, neurosurgery probably would not be where it is today. Harvey Cushing may never have introduced the bipolar cautery, and Gazi Yaşargil may not have delivered the operating microscope.

For this reason, some would argue that in certain circumstances, responsible innovation is actually a moral obligation of surgeons. The unique nature of surgery justifies its somewhat exceptional ethical and regulatory status, and applying the strict paradigm of systematic research and oversight—as is common for the introduction of new medicines—will stifle innovation. But even so, not every innovation will turn out to be an improvement over an existing procedure. For example, interspinous devices have been approved and extensively used before studies showed major unforeseen side effects.

Others may argue that surgeons should not be allowed to perform insufficiently validated interventions and that these procedures should only be accessible in the context of a systematic study evaluating their safety or efficacy. This implies that every innovation should be considered research and that the introduction of the novel procedure or device should follow the accepted principles of research ethics. A potential risk is that it could slow down innovation and severely limit access to novel (and beneficial) treatments.

Despite these opposing views, most would agree that innovation is fundamentally different from both standard care and research. This distinction is important as each has specific ethical, economic, administrative, and legal consequences, with particular responsibilities, rights, and obligations for doctors, hospitals, and other stakeholders.

According to The Belmont Report, a formative early guideline on clinical research ethics, innovative care is "practice that departs significantly from the standard or accepted." The intent of innovation is, unlike research, not to generate generalizable knowledge. At the same time, even though an innovative treatment aims to benefit the individual patient, because it is not a validated procedure, the risks will be unknown.

Innovation in neurosurgery comes with unique ethical challenges, which this book aims to highlight and explore. It is intended for neurosurgeons, neurosurgical residents, and medical students, but might be of interest to other surgeons, residents, and those involved in innovation as well. This book is divided into four parts.

Part I will detail the key ethical challenges of innovation in neurosurgery, including the lack of a clear definition of neurosurgical innovation, oversight and regulation, learning curves for innovative procedures, informed consent for innovation, innovation in children, and conflict of interest. Part II will discuss the payment for and right to innovative treatments and include the following questions: Can the public exert pressure for innovation? Do patients have a right to innovative treatments? And what is the role of hospital boards in innovative treatments? Part III covers the evaluation of neurosurgical innovations. The first of these chapters will examine the ethics of evaluating awake craniotomies for tumors—an example of a commonly performed and continuously evolving neurosurgical procedure. The remaining chapters will discuss specific ethical challenges related to recent trends in neurosurgery: personalized medicine and passive data collection. The last chapter will discuss ethical evaluation of a neurosurgical procedure that should be clearly considered research. Part IV closes this volume with a discussion about the teamwork and innovation followed by a chapter on the culture needed for innovation. Borrowing a few select topics from these chapters, the *Perspective* will provide a preliminary peek of the future of innovation in neurosurgery.

It is my sincere hope that the specific ethical challenges of innovation in neurosurgery present us with key opportunities to improve not only patient outcomes but also neurosurgery as a field.

Leiden, The Netherlands   Marike L. D. Broekman
The Hague, The Netherlands   On behalf of all coauthors

Acknowledgments

This book is the result of a collaborative effort of all contributors. I would like to thank each of them for their work, time, and support. It has been a true pleasure working with everyone. In addition, I would like to thank the team at Springer for their responsiveness and help.

Last and certainly not least, I would like to thank my family for their constant support.

Marike L. D. Broekman

Contents

Part I Innovation in Neurosurgery: Key Ethical Challenges

**1 Defining Innovation in Neurosurgery:
Challenges and Implications** 3
Mark M. Zaki, David J. Cote, and Marike L. D. Broekman

2 Informed Consent for Neurosurgical Innovation 11
Faith C. Robertson, Tiit Mathiesen, and Marike L. D. Broekman

**3 Ethical Challenges of Current Oversight
and Regulation of Novel Medical Devices in Neurosurgery** 27
Ivo S. Muskens, Saksham Gupta, Alexander F. C. Hulsbergen,
Wouter A. Moojen, and Marike L. D. Broekman

4 Ethics of Neurosurgical Innovation: Oversight and Regulation 39
Saksham Gupta, Ivo S. Muskens, Luis Bradley Fandino,
Alexander F. C. Hulsbergen, and Marike L. D. Broekman

5 The Ethics of the Learning Curve in Innovative Neurosurgery 49
Ludwike W. M. van Kalmthout, Ivo S. Muskens,
Joseph P. Castlen, Nayan Lamba, Marike L. D. Broekman,
and Annelien L. Bredenoord

6 Innovation in Pediatric Neurosurgery: The Ethical Agenda 57
Bart Lutters, Eelco Hoving, and Marike L. D. Broekman

7 Conflicts of Interest in Neurosurgical Innovation 65
Aislyn C. DiRisio, Ivo S. Muskens, David J. Cote,
William B. Gormley, Timothy R. Smith, Wouter A. Moojen,
and Marike L. D. Broekman

8 The Ethics of Funding Innovation: Who Should Pay? 75
Joseph P. Castlen, David J. Cote, and Marike L. D. Broekman

Part II   Payment for and Right to Innovation in Neurosurgery

9   Public Pressure for Neurosurgical Innovation 85
David J. Cote

**10   Surgical Innovation for Terminal Illnesses:
Do Patients Have a Right to Access Innovative Treatments?** 93
David J. Cote

**11   Ethics Committees, Innovative Surgery,
and Organizational Ethics** 105
Joseph P. Castlen and Thomas I. Cochrane

**12   Evaluating Awake Craniotomies in Glioma Patients:
Meeting the Challenge** 113
Bart Lutters and Marike L. D. Broekman

Part III   Evaluation of Innovations in Neurosurgery

**13   Ethical Considerations of Neuro-oncology Trial Design
in the Era of Precision Medicine** 121
Saksham Gupta, Timothy R. Smith, and Marike L. D. Broekman

**14   The Ethics of Passive Data and Digital Phenotyping
in Neurosurgery** 129
Joeky T. Senders, Nicole Maher, Alexander F. C. Hulsbergen,
Nayan Lamba, Annelien L. Bredenoord,
and Marike L. D. Broekman

15   Research Ethics: When Innovation Is Clearly Research 143
Nayan Lamba and Marike L. D. Broekman

**Part IV   Innovation in Neurosurgery: Required Culture and Team
Collaboration**

16   Innovation and Team Collaboration in Neurosurgery 153
Saskia M. Peerdeman

**17   Culture and Attitudes Supporting Ethical Innovation
in Neurosurgery** 159
Marjel van Dam and Marike L. D. Broekman

18   Perspective: Future of Innovation in Neurosurgery 165
Marike L. D. Broekman

Contributors

Marike L. D. Broekman, MD, PhD, LLM Department of Neurosurgery, Computational Neurosciences Outcomes Center (CNOC), Brigham and Women's Hospital, Harvard Medical School, Boston, MA, USA

Department of Neurosurgery, Haaglanden Medical Center, The Hague, The Netherlands

Department of Neurosurgery, Leiden University Medical Center, Leiden, The Netherlands

Annelien L. Bredenoord, PhD Department of Medical Humanities, Julius Center for Health Sciences and Primary Care, University Medical Center, Utrecht, The Netherlands

Joseph P. Castlen, BSc Department of Neurosurgery, Computational Neurosciences Outcomes Center (CNOC), Brigham and Women's Hospital, Harvard Medical School, Boston, MA, USA

Thomas I. Cochrane, MD, MBA Department of Neurology, Brigham and Women's Hospital and Center for Bioethics, Harvard Medical School, Boston, MA, USA

David J. Cote, MSc Department of Neurosurgery, Computational Neurosciences Outcomes Center (CNOC), Brigham and Women's Hospital, Harvard Medical School, Boston, MA, USA

Aislyn C. DiRisio, BSc Department of Neurosurgery, Computational Neurosciences Outcomes Center (CNOC), Brigham and Women's Hospital, Harvard Medical School, Boston, MA, USA

Luis Bradley Fandino, BSc Department of Neurosurgery, Computational Neurosciences Outcomes Center (CNOC), Brigham and Women's Hospital, Harvard Medical School, Boston, MA, USA

William B. Gormley, MD, PhD, MBA Department of Neurosurgery, Computational Neurosciences Outcomes Center (CNOC), Brigham and Women's Hospital, Harvard Medical School, Boston, MA, USA

Saksham Gupta, BA Department of Neurosurgery, Computational Neurosciences Outcomes Center (CNOC), Brigham and Women's Hospital, Harvard Medical School, Boston, MA, USA

Eelco Hoving, MD, PhD Department of Pediatric Neurosurgery, Brain Center Rudolf Magnus, University Medical Center Utrecht—Princess Máxima Center, Utrecht, The Netherlands

Alexander F. C. Hulsbergen, BSc Department of Neurosurgery, Computational Neurosciences Outcomes Center (CNOC), Brigham and Women's Hospital, Harvard Medical School, Boston, MA, USA

Department of Neurosurgery, Haaglanden Medical Center, The Hague, The Netherlands

Department of Neurosurgery, Leiden University Medical Center, Leiden, The Netherlands

Nayan Lamba, MSc Department of Neurosurgery, Computational Neurosciences Outcomes Center (CNOC), Brigham and Women's Hospital, Harvard Medical School, Boston, MA, USA

Bart Lutters, MD Department of Neurosurgery, Erasmus University Medical Center, Rotterdam, The Netherlands

Department of Pediatric Neurosurgery, Brain Center Rudolf Magnus, University Medical Center Utrecht—Princess Máxima Center, Utrecht, The Netherlands

Nicole Maher Department of Neurosurgery, Computational Neurosciences Outcomes Center (CNOC), Brigham and Women's Hospital, Harvard Medical School, Boston, MA, USA

Tiit Mathiesen, MD, PhD Department of Neurosurgery, Rigshospitalet and University of Copenhagen, Copenhagen, Denmark

Wouter A. Moojen, MD, PhD, MPH Department of Neurosurgery, Haaglanden Medical Center, The Hague, The Netherlands

Department of Neurosurgery, Leiden University Medical Center, Leiden, The Netherlands

Department of Neurosurgery, Haga Teaching Hospital, The Hague, The Netherlands

Ivo S. Muskens, MD Department of Neurosurgery, Haaglanden Medical Center, The Hague, The Netherlands

Department of Neurosurgery, Leiden University Medical Center, Leiden, The Netherlands

Department of Neurosurgery, Computational Neurosciences Outcomes Center (CNOC), Brigham and Women's Hospital, Harvard Medical School, Boston, MA, USA

Saskia M. Peerdeman, MD, PhD Department of Neurosurgery, Amsterdam University Medical Centers, Amsterdam, The Netherlands

Faith C. Robertson, BSc Department of Neurosurgery, Computational Neurosurgical Outcomes Center, Brigham and Women's Hospital, Harvard Medical School, Boston, MA, USA

Joeky T. Senders, MSc Department of Neurosurgery, Haaglanden Medical Center, The Hague, The Netherlands

Department of Neurosurgery, Leiden University Medical Center, Leiden, The Netherlands

Timothy R. Smith, MD, PhD, MPH Department of Neurosurgery, Computational Neurosciences Outcomes Center (CNOC), Brigham and Women's Hospital, Harvard Medical School, Boston, MA, USA

Marjel van Dam, MD, PhD Intensive Care Center, University Medical Center Utrecht, Utrecht, The Netherlands

Center for Research and Development of Education, University Medical Center Utrecht, Utrecht, The Netherlands

Ludwike W. M. van Kalmthout, MD Department of Medical Humanities, Julius Center for Health Sciences and Primary Care, University Medical Center, Utrecht, The Netherlands

Mark M. Zaki, BSc, BA Department of Neurosurgery, Computational Neurosciences Outcomes Center (CNOC), Brigham and Women's Hospital, Harvard Medical School, Boston, MA, USA

Abbreviations

5-ALA	5-Aminolevulinic acid
CCTI	Capacity to Consent to Treatment Instrument
CE	Conformité Européenne
CEO	Chief executive officer
COIs	Conflicts of interest
CRT	Cluster randomized trial
CSRF	Cushing's Support and Research Foundation
DBS	Deep brain stimulation
EANS	European Association of Neurosurgical Societies
EC	Ethics committee
EEA	European Economic Area
EEG	Electroencephalogram
EEMS	Endoscopic endonasal approach for skull base meningiomas
EU	European Union
EUDAMED	European Database on Medical Devices
FDA	Food and Drug Administration
GBM	Glioblastoma
I(P)Ds	Interspinous (process) devices
ICU	Intensive care unit
IDE	Investigational device exemption
IDEAL	Idea, Development, Exploration, Assessment, Long-term Follow-up
IDEAL-D	Idea, Development, Exploration, Assessment, Long-term Follow-up Device
IECs	Institutional ethics committees
IRB	Internal review board
JCAHO	Joint Commission on Accreditation of Healthcare Organizations
LHS	Learning health system
MacCAT-T	MacArthur Competence Assessment Tool for Treatment
MAUDE	Manufacturer and User Facility Device Experience
MDM	Multidisciplinary team meetings
MedSun	Medical Product Safety Network
MHRA	Medicines and Healthcare Products Regulatory Agency
MSIIT	Macquarie Surgical Innovation Identification Tool
NGS	Next-generation sequencing

NIH	National Institutes of Health
OTC	Observational treatment comparisons
PD	Passive data
PMA	Pre-market approval
PPACA	Patient Protection and Affordable Care Act
QI	Quality improvement
R&D	Research and development
RCT	Randomized controlled trial
rhBMP	Recombinant human bone morphogenetic protein
SES	Socioeconomic status
SIC	Surgical Innovation Committee
USA	United States of America
WEB	Woven EndoBridge
WFNS	World Federation of Neurosurgical Societies

Part I

Innovation in Neurosurgery: Key Ethical Challenges

Defining Innovation in Neurosurgery: Challenges and Implications

Mark M. Zaki, David J. Cote, and Marike L. D. Broekman

Introduction

Innovation is an important part of the practice of neurosurgery. In a continually evolving field, neurosurgeons must frequently assess and reassess the most appropriate and effective treatments for each patient. Innovation is conducted by neurosurgeons in a range of settings, from those investigating novel treatments for brain tumors in major academic institutions to those performing creative surgeries in low-resource settings across the world [1, 2]. Yet, something so ubiquitous among neurosurgeons remains difficult to define with consensus. To this day, great heterogeneity exists in what surgeons consider innovative [3, 4]. Different interpretations of what constitutes innovation lead to a lack of standardization in evaluating novel procedures across surgeons, departments, institutions, and nations. Additionally, proof of the innovative nature of a project is often a key component of securing grant funding; therefore, efforts to standardize what should be considered innovative could be beneficial to funding agencies.

This chapter is in part based on: Zaki MM, Cote DJ, Muskens IS, Smith TR, Broekman ML: Defining Innovation in Neurosurgery: Results from an International Survey. **World Neurosurg**, March 29, 2018.

M. M. Zaki · D. J. Cote
Department of Neurosurgery, Computational Neurosciences Outcomes Center (CNOC), Brigham and Women's Hospital, Harvard Medical School, Boston, MA, USA

M. L. D. Broekman (✉)
Department of Neurosurgery, Computational Neurosciences Outcomes Center (CNOC), Brigham and Women's Hospital, Harvard Medical School, Boston, MA, USA

Department of Neurosurgery, Haaglanden Medical Center, The Hague, The Netherlands

Department of Neurosurgery, Leiden University Medical Center, Leiden, The Netherlands
e-mail: m.broekman@haaglandenmc.nl; m.l.d.broekman@lumc.nl

Attempts to standardize the definition of innovation in the surgical literature have been presented. The Society of University Surgeons has proposed discerning between variations, innovations, and research [5]. Some have suggested splitting innovations by type, such as minor modifications of standard procedures, major modifications of standard procedures, and innovations that are new to the institution but have been validated elsewhere [6]. Others have suggested a rating of surgical innovations directly related to the amount of oversight deemed necessary [7]. Despite these attempts, along with many other suggestions for appropriate oversight in surgery [8–16], a clear answer does not exist.

Consistent with the general surgical literature [17–19], neurosurgeons also show great heterogeneity in what they consider to be an innovative procedure. For example, an international survey showed that 47% of neurosurgeons consider the use of a new high-speed drill for a transsphenoidal approach to be innovative—nearly an equal divide in opinion among the approximately 350 respondents. The same survey indicated that even when there was a general consensus whether a procedure was considered innovative, there was disagreement concerning whether oversight was necessary. For example, although 79% of respondents considered the use of a new dura substitute to be innovative, only 17% suggested that ethical standards require prior approval before its use [20]. Such diversity of thought likely reflects the complexity in identifying an innovative procedure and determining the amount of oversight deemed necessary. In this chapter—which is partially based on our recent survey [20]—we discuss the current challenges in achieving a unified understanding of innovation in neurosurgery and the ethical problems that arise in the absence of a clear understanding of innovation in neurosurgery. Finally, we discuss possible solutions for uniting the field moving forward.

The Difficulty in Defining Innovation in Neurosurgery

Innovations are not unique to neurosurgery or to medicine. In the business literature, innovations can broadly be categorized into sustaining and disruptive innovations [21]. Sustaining innovations improve an existing product and maintain the incumbent industry. One example can be the latest version of an existing smartphone. Disruptive innovations introduce a new product that radically disturbs an existing industry, such as the effect of Uber on the taxi industry [22]. Businesses can predict the type of innovation a particular product will be by using consumer reports and market predictions to guide the development and marketing of their products. Surgery, however, is not driven primarily by consumer requests and other market forces. It is instead guided by surgeon preference, patient outcomes, and peer review [23].

Innovation in surgery varies drastically even from other fields of medicine [24]. Medical innovations, such as devices or new drugs, undergo a rigorous and thorough evaluation before they are approved for the clinical market. Once they are introduced, these innovations are believed to be safe and effective in achieving the desired effect. Thus, there is a clear border between research and clinical care in medicine. Surgery is more complicated. Since the Food and Drug Administration or an equivalent organization does not typically review the safety and efficacy of new surgical procedures [25], it is noteworthy that research and clinical care are not mutually exclusive in surgical hypothesis testing. Surgical innovation in both the research and the clinical paradigm may contain untested novel ideas [7], but innovation in research is aimed at generating generalizable knowledge [26, 27], while innovation in clinical care is aimed at improving the outcome of the individual patient [26]. When new surgical procedures are implemented in patients, generating universal knowledge thus coincides with the aim of ameliorating the suffering of the individual patient. Such overlap, along with the lack of oversight, has obfuscated a clear definition of innovation in the surgical field.

Ethical Implications: The Need for Consensus About What Constitutes Innovation

Ethically, physicians are called to do no harm. Rapid application before proper evaluation has historically led to compromising patient safety. For example, the widespread use of frontal lobotomy before it was properly evaluated led to numerous undesired consequences [28]. Being able to a priori define what may constitute innovation would thus ensure appropriate evaluation of patient safety and ethical care before implementing an innovation into clinical practice.

Often, the person introducing the innovation is the surgeon using the novel technique or device. In scenarios where the surgeon is the one who strongly believes in the promise of the innovation, innovator bias may prevent the surgeon from thoroughly evaluating the potential harms associated with the new intervention [29]. Such lack of perceived clinical equipoise and other personal conflicts of interests are therefore just as important to be aware of as financial conflicts of interest [30]. Non-biased evaluation may help to limit the effect of such conflicts of interest in cases where a new idea is clearly defined as an innovation.

Furthermore, the principle of patient autonomy is contingent upon informed consent [31], and it is controversial whether or not the consent patients provide for new surgeries is truly informed [32]. A key component of informed consent is that the relevant risks and benefits are disclosed to the patient, as well as the details of the procedure itself. If a new innovation is being implemented, in which the risks are unknown, the patient may not be able to offer truly appropriate informed consent [33]. Even if certain patients tend to put full trust in their surgeon without knowing all of the details of the procedure [24], it is important that all relevant information be available to the patient and the surgeon in order to make an informed decision plan. Knowing when to critically evaluate a novel innovation and when to directly make an insignificant change, such as using a new type of suture, depends on how innovation is defined.

The principle of justice can also be explored, both in regard to over-enrolling vulnerable patient populations as well as under-enrolling patients from disadvantaged backgrounds. Because severely ill neurosurgical patients may not have the cognitive ability to adjudicate risks and benefits, and because they and their caretakers may have a strong emotional drive to attempt any option feasible, these patients are susceptible to being easily persuaded into a novel treatment [34]. Regarding under-access, minority and low-income neuro-oncology patients have worse access to surgical care than Caucasian patients or those who have higher incomes, respectively [35]. Since many new innovations tend to be costly, low-income patients may not benefit from potentially lifesaving treatments [36]. Conversely, dangerous innovations may be forced onto minority populations as has occurred in Tuskagee [37]. Without a proper framework of innovation or appropriate oversight, these injustices are prone to exacerbation.

As seen above, surgeon innovators should take into account certain ethical principles to ensure appropriate patient safety. It is important to assess both financial and personal conflicts of interest when determining the value of an innovation. Patient autonomy must be upheld, and both the surgeon and patient should be as informed as possible about all potential consequences. Finally, patient selection should be carefully performed to prevent exacerbating current injustices associated with access to surgical care. Potentially compromising these core ethical principles thus necessitates a standard definition of innovation in order to ensure ethical and practical patient safety.

Lack of a Definition: Implications for Oversight

As indicated above, neurosurgeons do not agree on what constitutes innovation, which has contributed to the disagreement in what is deemed appropriate regulation for novel procedures. Some fear that oversight may stifle innovation and the continual advancement of surgery [38]; however, appropriate oversight that balances patient safety and the surgeon's autonomy is the goal. Many proposals have been suggested for achieving appropriate oversight in cases where deviations from the norm take place, whether they are technical or technological deviations. We have previously reviewed [6] the proposals for various types of innovations, including those that suggest national regulation for major modifications or radical innovations [39] as well as those that suggest an institutional surgical innovation committee (SIC) [29, 40, 41].

Irrespective of the type of regulation, the first step in determining what oversight is appropriate for an innovative surgery is to determine which operations require an evaluation in the first place. On one extreme, every operation may be considered a deviation from the norm, as surgeons tailor their operations to the uniqueness of each presentation [23]. It would of course be impractical and inefficient to evaluate every deviation from the norm, however. A different approach would be to introduce a checklist that can help identify if a procedure qualifies as innovative and needs evaluation. To this aim, the Macquarie Surgical Innovation Identification Checklist (MSIIT) has been introduced and is currently being tested in its ability to identify innovations in the clinic [42].

The value of the MSIIT is not that it seeks to create stringent criteria for what constitutes innovation but is helpful in regard to its ability to identify which surgical procedures warrant further information and oversight when necessary.

After identifying which procedures warrant further oversight, the regulation that is deemed appropriate could be determined by an SIC. An SIC may be comprised of experienced surgeons, ethicists, engineers, and other relevant stakeholders. When a new technique or device is being introduced, the SIC can critically evaluate the scientific validity of the proposal, ensure that the patient is truly informed about all known risks and the novelty of the procedure, and confirm that the necessary adaptations to the novel procedure are made available to the surgeon. When an innovation has been proven effective, there will be a learning curve that must be overcome before other colleagues are able to effectively incorporate it into common practice [43, 44]. SICs can serve as facilitators that connect experienced surgeons with similar ideas and experiences to foster educational dialogue between colleagues.

Conclusion and Future Directions

Overall, surgical innovation is a ubiquitous phenomenon that remains poorly defined. This lack of consensus poses practical and ethical concerns relevant to appropriate oversight of innovative procedures. As it is impractical to evaluate every deviation from the norm, checklists to identify when a procedure qualifies as innovative might form a pragmatic solution so that neurosurgeons can continue to advance the field of neurosurgery without compromising patient safety or the fundamental ethical principles of medical practice.

References

1. Bean JR. Neurosurgical innovation in the developing world: where will it come from? World Neurosurg. 2015;84:1522–4.
2. Ravindra VM, Kraus KL, Riva-Cambrin JK, Kestle JR. The need for cost-effective neurosurgical innovation—a global surgery initiative. World Neurosurg. 2015;84:1458–61.
3. Reitsma AM, Moreno JD. Ethics of innovative surgery: US surgeons' definitions, knowledge, and attitudes. J Am Coll Surg. 2005;200:103–10.
4. Rogers WA, Lotz M, Hutchison K, Pourmoslemi A, Eyers A. Identifying surgical innovation: a qualitative study of surgeons' views. Ann Surg. 2014;259:273–8.
5. Biffl WL, Spain DA, Reitsma AM, Minter RM, Upperman J, Wilson M, et al. Responsible development and application of surgical innovations: a position statement of the Society of University Surgeons. J Am Coll Surg. 2008;206:1204–9.
6. Broekman ML, Carriere ME, Bredenoord AL. Surgical innovation: the ethical agenda: a systematic review. Medicine (Baltimore). 2016;95:e3790.
7. Bernstein M, Bampoe J. Surgical innovation or surgical evolution: an ethical and practical guide to handling novel neurosurgical procedures. J Neurosurg. 2004;100:2–7.
8. Awad IA. Innovation through minimalism: assessing emerging technology in neurosurgery. Clin Neurosurg. 1996;43:303–16.
9. Beech R, Morgan M. Constraints on innovatory practice: the case of day surgery in the NHS. Int J Health Plann Manag. 1992;7:133–48.
10. Bunker JP, Hinkley D, McDermott WV. Surgical innovation and its evaluation. Science. 1978;200:937–41.
11. Ergina PL, Cook JA, Blazeby JM, Boutron I, Clavien PA, Reeves BC, et al. Challenges in evaluating surgical innovation. Lancet. 2009;374:1097–104.

12. Frader JE, Flanagan-Klygis E. Innovation and research in pediatric surgery. Semin Pediatr Surg. 2001;10:198–203.
13. Gillett G. Ethics of surgical innovation. Br J Surg. 2001;88:897–8.
14. Haines SJ. Randomized clinical trials in the evaluation of surgical innovation. J Neurosurg. 1979;51:5–11.
15. McKneally MF. Ethical problems in surgery: innovation leading to unforeseen complications. World J Surg. 1999;23:786–8.
16. Schwartz JA. Innovation in pediatric surgery: the surgical innovation continuum and the ETHICAL model. J Pediatr Surg. 2014;49:639–45.
17. ACOG Committee on Ethics. ACOG Committee Opinion No. 352: innovative practice: ethical guidelines. Obstet Gynecol. 2006;108:1589–95.
18. Lieberman JR, Wenger N. New technology and the orthopaedic surgeon: are you protecting your patients? Clin Orthop Relat Res. 2004;(429):338–41.
19. Wang Y, Kotsis SV, Chung KC. Applying the concepts of innovation strategies to plastic surgery. Plast Reconstr Surg. 2013;132:483–90.
20. Zaki MM, Cote DJ, Muskens IS, Smith TR, Broekman ML. Defining innovation in neurosurgery: results from an international survey. World Neurosurg. 2018;114:e1038–48.
21. Hwang J, Christensen CM. Disruptive innovation in health care delivery: a framework for business-model innovation. Health Aff (Millwood). 2008;27:1329–35.
22. Schneider H. Creative destruction and the sharing economy: Uber as disruptive innovation. Cheltenham: Edward Elgar Publishing; 2017.
23. Riskin DJ, Longaker MT, Gertner M, Krummel TM. Innovation in surgery: a historical perspective. Ann Surg. 2006;244:686–93.
24. Angelos P. Surgical ethics and the challenge of surgical innovation. Am J Surg. 2014;208:881–5.
25. Spodick DH. Numerators without denominators. There is no FDA for the surgeon. JAMA. 1975;232:35–6.
26. United States. National Commission for the Protection of Human Subjects of Biomedical and Behavioral Research. The Belmont report: ethical principles and guidelines for the protection of human subjects of research. Bethesda: The Commission; 1978.
27. World Medical Association. World Medical Association Declaration of Helsinki: ethical principles for medical research involving human subjects. JAMA. 2013;310:2191–4.
28. Agarwal P, Sarris CE, Herschman Y, Agarwal N, Mammis A. Schizophrenia and neurosurgery: a dark past with hope of a brighter future. J Clin Neurosci. 2016;34:53–8.
29. Morreim H, Mack MJ, Sade RM. Surgical innovation: too risky to remain unregulated? Ann Thorac Surg. 2006;82:1957–65.
30. Minter RM, Angelos P, Coimbra R, Dale P, de Vera ME, Hardacre J, et al. Ethical management of conflict of interest: proposed standards for academic surgical societies. J Am Coll Surg. 2011;213:677–82.
31. Hall DE, Prochazka AV, Fink AS. Informed consent for clinical treatment. CMAJ. 2012;184:533–40.
32. Falagas ME, Korbila IP, Giannopoulou KP, Kondilis BK, Peppas G. Informed consent: how much and what do patients understand? Am J Surg. 2009;198:420–35.
33. Geiger JD, Hirschl RB. Innovation in surgical technology and techniques: challenges and ethical issues. Semin Pediatr Surg. 2015;24:115–21.
34. Ford PJ. Vulnerable brains: research ethics and neurosurgical patients. J Law Med Ethics. 2009;37:73–82.
35. Mukherjee D, Zaidi HA, Kosztowski T, Chaichana KL, Brem H, Chang DC, et al. Disparities in access to neuro-oncologic care in the United States. Arch Surg. 2010;145:247–53.
36. Hutchison K, Johnson J, Carter D. Justice and surgical innovation: the case of robotic prostatectomy. Bioethics. 2016;30:536–46.
37. Howell J. Race and U.S. medical experimentation: the case of Tuskegee. Cad Saude Publica. 2017;33(Suppl 1):e00168016.
38. Moojen WA, Bredenoord AL, Viergever RF, Peul WC. Scientific evaluation of spinal implants: an ethical necessity. Spine (Phila Pa 1976). 2014;39:2115–8.

39. Qualms about innovative surgery. Lancet. 1985;1:149.
40. Georgeson K. Surgical innovation and quality assurance: can we have both? Semin Pediatr Surg. 2015;24:112–4.
41. McKneally MF, Daar AS. Introducing new technologies: protecting subjects of surgical innovation and research. World J Surg. 2003;27:930–4.; ; discussion 934–935.
42. Blakely B, Selwood A, Rogers WA, Clay-Williams R. Macquarie Surgical Innovation Identification Tool (MSIIT): a study protocol for a usability and pilot test. BMJ Open. 2016;6:e013704.
43. Cook JA. The challenges faced in the design, conduct and analysis of surgical randomised controlled trials. Trials. 2009;10:9.
44. Mayer E, Darzi A. Innovation and surgical clinical trials. Lancet. 2016;388:1027–8.

Informed Consent for Neurosurgical Innovation

Faith C. Robertson, Tiit Mathiesen, and Marike L. D. Broekman

Introduction

The spectrum of innovation within neurosurgery ranges from tactical delivery of experimental molecular therapies and anatomical use of 3D-printed biosynthetic structures [1, 2] to novel approaches in surgical planning and healthcare system strengthening [3, 4]. These advances propel the field forward, making patient care possible in ways previously unfathomed. However, when integrating innovation into clinical practice, there are ethical challenges distinct from both research and clinical care, particularly regarding a patient's understanding of what novel treatment entails. Many candidates for innovative treatments have illnesses refractory to standard therapies, and their reliance on new alternatives supersedes the demand for more substantial evidence prior to treatment. In contrast, for cases of procedural modification and new technologies, where the agreement to undergo an innovative surgery is more elective, there may be misconceptions about what the new approach entails or promises or how many times it has been previously tried. Furthermore, there exists subtle ambiguity between what is patient-specific problem solving and advancing the field

F. C. Robertson
Department of Neurosurgery, Computational Neurosurgical Outcomes Center, Brigham and Women's Hospital, Harvard Medical School, Boston, MA, USA

T. Mathiesen
Department of Neurosurgery, Rigshospitalet and University of Copenhagen, Copenhagen, Denmark

M. L. D. Broekman (✉)
Department of Neurosurgery, Computational Neurosciences Outcomes Center, Brigham and Women's Hospital, Harvard Medical School, Boston, MA, USA

Department of Neurosurgery, Haaglanden Medical Center, The Hague, The Netherlands

Department of Neurosurgery, Leiden University Medical Center, Leiden, The Netherlands
e-mail: m.broekman@haaglandenmc.nl; m.l.d.broekman@lumc.nl

© Springer Nature Switzerland AG 2019
M. L. D. Broekman (ed.), *Ethics of Innovation in Neurosurgery*,
https://doi.org/10.1007/978-3-030-05502-8_2

through experimental research. A surgeon's care for a patient is generally directed toward the present health and welfare of the individual, whereas clinical investigators act to generate knowledge that may contribute to therapeutic benefits for future patients; with an innovative method, the clinician and investigator may be one in the same. Therefore, a delicate approach is required when translating innovation to the operating room, especially regarding the process of informed consent.

Informed consent is a critical medical, legal, and ethical requirement of physicians and surgeons prior to initiating a treatment plan [5]. The process relies on appropriate provision of information to a competent patient in efforts to permit patient autonomy over healthcare decision-making without coercion. Importantly, informed consent is not isolated to a single conversation and document signing but is rather an ongoing process of communication throughout the trajectory of the patient's care. However, multiple factors, including illness itself, influence a patient's ability to make his or her own decisions with true autonomy.

In neurosurgery, the execution of informed consent is further complicated; neurosurgical patients are one of the most vulnerable populations. First, for disease processes affecting information processing or the ability to participate in high-level cognitive decision-making, an individual's capacity to partake in informed consent may be hindered [6–9]. Second, the content provided to the patient must be sufficient to formulate a knowledgeable decision. However, in the advent of innovation and novel approaches to treatment, there is tremendous debate and variability in practice surrounding the content or detail of information that should be disclosed. Finally, patient consent must be free from coercion, that is, from the individual's personal community or the medical community. Coercion avoidance is especially pertinent in the setting of innovative therapies, as the medical community must not exaggerate the benefits of a recommended treatment, exaggerate harm of choosing an alternative, strategically add or remove barriers to the individual, or proceed with a greater intent to benefit a research agenda than the patient at hand.

Overall, innovation in neurosurgery presents ethical challenges regarding consent, as innovative treatment options may provide better patient outcomes, but unprecedented surgical interventions may include unknown risk. This review summarizes the importance and difficulties of informed consent within neurosurgery, including patient capacity, content and format of discussion, and coercion—all key factors in the attainment of proper consent and the clinical decision process. Ultimately, we underscore the inherent complexity in balancing scientific evidence, clinical expertise, and patient and family preference when pursuing innovative neurosurgical treatments, in efforts to bring about discussion on improvements we can make within the field.

Capacity

Fundamentally, to participate in informed consent, patients must have the mental ability to communicate a choice, understand the relevant information, appreciate the medical consequences of the choices at hand, and express reason for their decision

[6]. However, many disease processes afflicting the neurosurgical population inhibit this decision-making ability, particularly later in the disease course when symptoms are refractory to standard treatment and the patient is now seeking innovative methods. For instance, a 2014 study by Kerrigan et al. on 100 patients with intracranial tumors demonstrated cognitive impairment in 25% of patients and 38% of patients with grade IV pathology [10]. Such high statistics were also noted in 2009 in a cross-sectional study that demonstrated more than 50% of the patients with malignant gliomas had impaired capacity [11]. This tumor-associated mental incapacity remains largely under-recognized by physicians and has been attributed to insufficient preoperative assessment [10, 12–15]. Apart from malignancy, many other neurological diseases hold concomitant cognitive challenges. In a study of 20 patients with Parkinson's disease, 20% were unable to evidence choice of treatment, and 80% were unable to understand treatment [16]. Limitations in patients' capacity are also well documented in psychopathology and trauma [17–22].

Failure to recognize the patient's limitations and need for a surrogate decision-maker is a serious clinical problem, as it allows patients to be guided by their surgeon toward a particular option without adequate comprehension of the decision. Physicians must become proficient in evaluating the patient's capacity and executing proper informed consent [6]. To better recognize when additional counseling or recruitment of a healthcare proxy may be required, formal preoperative cognitive assessments such as the MacArthur Competence Assessment Tool for Treatment (MacCAT-T) and the Capacity to Consent to Treatment Instrument (CCTI) have been suggested [10–13, 16]. Alternative assessments like IQ measurement or minimental state examinations are less helpful in this context [15, 23]. In prior neurological assessment studies, the most critical and cognitively demanding consent standard was "understanding" or the ability to encode, consolidate, and recall relatively complex information [7–9, 11, 16]. Though, as time is often a limiting factor for clinicians, a rapid evaluation tool is desired. Triebel et al. used the CCTI and found that the primary cognitive predictor of understanding was a measure of short-term verbal acquisition and recall, particularly testing semantic fluency—having patients recall as many animals as possible from a verbalized list in 1 min—on the Hopkins Verbal Learning Test [11]. As "reasoning" appears to be second to understanding with respect to cognitive demand [7, 8, 24], others have proposed formulas to assess a patients' logic, episodic memory, and processing speed to determine their ability to decide on treatment options [25]. While it is evident that more stringent cognitive testing is needed in the clinical setting, the question of whether a specific neuropsychological disability should define a patient's decisional capacity for treatment remains [26].

If the patient is deemed to not have capacity, what are the ethics behind treating these individuals with innovative or experimental therapy? The ethical requisites for research involving persons with mental derangement fall into two classes, protection versus access [27, 28]. Strict protection aims to protect subjects from risks of harm, even at the cost of slowing the progress of scientific investigation, the development of new medical advances, and the access to potentially beneficial but unproven therapies. Access theory advocates for greater admittance to participation in research protocols when standard therapy is inadequate. If the patient's capacity is lacking, decisions will progress through the hierarchy of clinical and legal standards [29], with the highest priority given to enacting the patient's previously expressed wishes (written or verbal). If the patient's wishes are unknown, the principle of substituted judgment via a surrogate decision-maker advises on what the patient would have wanted. Care should be taken in emergency scenarios, as it is not uncommon to confused proxies in conditions requiring emergent intervention [30]. If no information about the patient's values and preferences is available, decisions should be made according to the best interest standard [31].

Content

Adequate disclosure and delivery of information during the consent process are challenging at baseline. To assess how well patients were being consented for neurosurgery, irrespective of innovation, researchers at the National Hospital for Neurology and Neurosurgery in London conducted a prospective study on 60 patients undergoing a variety of neurosurgical procedures, and 97% of the patients felt that they had reached an informed decision [32]. However, only 25% reported being informed about the general risks of surgery and anesthesia, and only 33% felt fully informed of alternative treatment options. At the German Brain Tumor Centre, the complexity of consent forms was shown to negate its utility as a decision-making aid [33]. While all patient consent forms fulfilled ethical and legal requirements, graduate levels were required to read and understand five of nine documents, and forms were lacking in scientific evidence, social aspects, text length, formal layout, and readability. These challenges in the content relayed in informed consent are only further complicated when discussing innovation.

In informed consent discussions surrounding novel approaches to treatment, there is tremendous debate and variability in practice surrounding the content or detail of information that should be disclosed. The content of discussion must be sufficient to formulate a knowledgeable decision; however, there is no standard within surgery regarding the extent to which a surgeon should discuss the innovative nature of the procedure [14, 34–52], the evidence or lack thereof [47, 53–58], the associated risks and benefits [34, 37, 38, 42, 44–46, 49–51, 54, 56, 59–63], the unforeseeable or unknown risks given the experimental and invalidated nature of the procedure [38, 57], the operating surgeon's learning curve considering his or her experience with the procedure [34, 55–57, 64–67], and the alternative treatment options [37, 38, 40, 46, 49–51, 54, 59–63].

Paucity of Data

Whether a new treatment is superior to the standard is usually learned via clinical research, and while the introduction of novel pharmaceuticals progresses through a thorough, systematic process of regulations from bench to bedside, innovation in surgery often occurs without controlled study conditions. Randomized, controlled trials remain the gold standard of class I evidence and provide the most internally valid form of clinical evidence for treatment efficacy. However, this format of research is difficult to execute in surgery, as there is felt to be a lack of equipoise, and there are ethical considerations arising from sham surgery as a potential control within the double-blinded design [68, 69]. Furthermore, the results of surgical techniques are more tenuous to measure, as they may be operator dependent, and occur less frequently in time, resulting in an inability to accrue sufficient statistical power to inform future practice. This implicit paucity of prospective data makes it difficult to advise patients on predicted outcomes and inhibits their ability to comprehend the gravity of the intervention.

During the period in which outcomes data from an innovative procedure is limited, care must be taken in the discussion of associated risk. In one regard, the paucity of adverse events data may result in a higher proportion of patient agreement to innovative procedures. For instance, a study of consent before and after knowing the prognostic outcomes of a decompressive hemicraniectomy showed that participants' willingness to consent to the procedure decreased after learning of predicted risk [70]. In contrast, some believe that in-depth discussions that emphasize all potential unfavorable outcomes and adverse events, even when rare or uncertain, are likely to reframe the patients' expectations and may deleteriously influence both willingness to consent and patient outcomes. There is a well-documented "nocebo effect," or negative placebo effect, that can occur when patients are given all possible negative outcomes during the informed consent process; their reported experience of adverse events mimics the information given [71]. These nocebo effects have led between 4 and 26% of patients in placebo arms of clinical trials to withdraw from treatment secondary to perceived adverse events [71–73]. Meanwhile, another study on patients' recollection of risk associated with neurosurgical procedures argued that patients pay little attention to associated risks [74]. Authors showed at 2 h after an informed consent consultation, participants recalled only four risks out of a possible 32 for cranial surgery and 25 for spinal surgery (other procedure details were recalled more accurately). Thus, whether more prognostic data would change patient's treatment decision remains uncertain.

This theory of pre-treatment expectations translating to outcomes likely affects the level of detail relayed during the consent process, and may make surgeons less likely to disclose an expansive list of risks, particularly if there is limited data. However, it is advised that the current body of knowledge at the time of consultation should be conveyed to the patient in a balanced way, as to avoid creating either an overt sense of doom or false hope [75].

Details of Procedure

While the Society of University Surgeons has published guidelines on how operative innovations should be introduced and that informed consent "should include discussion of the innovative aspect of the procedure" [37], technical details of a procedure may be omitted due to time of the consultation or beliefs that the patient does not need to understand the granular details of surgery. Lee and colleagues conducted a patient and attending survey to understand what information patients and surgeons consider essential to disclose before an innovative surgical procedure [44]. Nearly 60% of patients said they could not commit to surgery without a general description of the procedure, whereas only 20% of surgeons considered it essential to discuss the technical details. Over 70% of these patients desired to understand known risks and benefits and acknowledgement of potentially unknown risks and benefits. In a separate study involving 25 patients who had undergone surgery for a benign brain tumor, arteriovenous malformation, or unruptured aneurysm, semi-structured interviews revealed a strong patient desire for procedural details [76]. Patients wanted to know exactly where the lesion was, the surgical entry point location, the likely size and shape of incision, which important nerves were in the vicinity, and how their skull would be reconstructed afterward. When this information was conveyed, patients reported a reduction in preoperative anxiety. As the desired information on technical detail varies between patients, it is recommended that the surgeon relay what is unique about this specific innovative approach, compared to standard treatment options, and the level of detail thereafter is at the discretion of the surgeon.

Technical Learning Curve

The concept of the learning curve, first identified in the 1880s, reflects that repetitions in tasks over time correspond with acquisition of a skill and optimal performance outcomes [77]. Within surgery, the learning curve must be acknowledged as a variable in the safety and efficacy of an operation [78, 79]. Estimates of cases required to achieve proficiency and risk reduction vary. Case-related proficiency—as indicated by notable improvement in outcomes and workflow for both the surgeon and anesthesia team—has been estimated to be approximately 40 cases for a minimally invasive transforaminal lumbar interbody fusion [80], between 17 and 100 cases in endoscopic transsphenoidal surgery [81, 82], and 25 cases of awake craniotomies [83]. For an innovative procedure, neurosurgeons are often on the ascending limb of the learning curve and the skill set of the surgeon and the technical difficulty of surgery will impact outcome. Therefore, enumeration of the surgeon's experience will likely correlate with success; however, this information is not always disclosed to the patient.

The paramount legal case in neurosurgery that addresses disclosure of experience as a risk factor in informed consent for neurosurgery is the case of *Johnson v. Kokemoor* [84, 85]. In this case, patient Johnson was diagnosed with an enlarging basilar aneurysm and was operated on by neurosurgeon Richard Kokemoor. The operation left her incontinent and unable to walk, with mild impairment of vision, speech, and upper body coordination. In court, she claimed she was deceived by the neurosurgeon's exaggeration of both her need for surgery and his experience in clipping basilar bifurcating aneurysms. When Johnson originally inquired about

Kokemoor's previous experience, he relayed that he had performed the surgery "several" times, and when she inquired further, Kokemoor indicated "dozens" and "lots of times." Kokemoor had only performed two basilar bifurcation aneurysms, neither as large or complicated as Johnson's case. Kokemoor was found guilty of assault in violating his informed consent obligations by failing to describe accurately his relative lack of experience with the surgery in question.

The aforementioned case brings forth the constant struggle surgeons face when trying to acquire more experience and skill in a new technique, the balance between patient-centered obligations and disclosure of inexperience, with desire to practice and improve technical skill for future patients. Had Dr. Kokemoor revealed how few procedures he had done, Ms. Johnson would have likely sought another surgeon. In a recent neurosurgical patient survey, the most desired information, regardless of innovation, was whether the surgeon was performing the procedure for the first time [44]. Additionally, over 70% of patients indicated that they could not decide on surgery without knowing the surgeon's training experience. In contrast, 20% fewer surgeons thought it was critical to discuss other surgeons' volumes and outcomes, the surgeon's own volumes and outcomes, and the surgeon's special training. Paradoxically, disclosure of inexperience may deter patients and potentiate continued inexperience of the surgeon. Nonetheless, ethically, one is still required to provide an honest retort.

Alternative Treatments

In order for patients to have true autonomy over their treatment decision, they must be aware of alternatives. However, even patients undergoing standard neurosurgical procedures often lack that understanding. In an English study in 2000, two-thirds of patients undergoing elective neurosurgery reported unawareness of alternative treatment options [32]. With innovative treatments, patients' recognition for alternatives becomes even more skewed, as there is common misconception that an "innovative" and "new" option is superior and constitutes improved treatment. Furthermore, patients who are being considered for innovative treatment options often have illnesses refractory to standard therapies, and the alternative treatments may be limited. These patients may feel pressured to try whatever is offered, and may not realize alternatives to the innovative procedure that is being premiered at the present institution, including the attempt for palliation as opposed to the attempt for cure. Neurosurgeons should also be mindful that patients may consent to avoid conflict or being perceived as adding negativity to the patient-doctor relationship [14, 86].

To mitigate bias in treatment recommendation, many academic institutions employ a multidisciplinary team that includes specialists from neurosurgery, neurology, pathology, oncology, and radiation, as well as ancillary staff, to explain the risks and benefits of various treatment options [3, 87, 88]. The process is designed to allow the patient a balanced presentation of information to foster a timely decision that ensures adequate informed consent. Many authors believe a similar approach should be taken when innovative options are posed, where consent to the innovation occurs after meeting with a larger team, and a significant portion of team members have a treatment role different from the surgeon. For patients being evaluated for innovative treatment near the end of life, it is advised that they are both well informed and some authors suggest seeking an independent surgeon for a second opinion [89].

Coercion

A patient's perspective on innovative treatment can be greatly influenced by terminology, misconceived notions about outcome, the patient-doctor relationship, and financial barriers or lack thereof. Persuasive language often goes unnoticed, but a surgeon's description of the surgery as "new" or "innovative" can generate misconceptions that innovations are improvement in practice that implicitly result in better outcomes. On innovation, McKneally et al. note: "the term *innovative* has a seductive connotation of added value, especially in a progressive society" [90]. To counter this subtle persuasion, multiple authors have recommended changing the vocabulary by describing new procedures to patients and the public as "non-validated" as opposed to "innovative."

The patient-doctor relationship introduces another level of coercion. A surgeon may emphasize an innovative treatment over another in benevolence, based on hope or desperation that a patient in which they are emotionally invested could benefit, even if the patient is not a proper candidate for the procedure. This can foster a shared sentiment of courage in commitment to the innovated procedure [91]. Additionally, the patient may trust the innovator as the "expert in the field," and the ensuing relationship takes on a more paternalistic dynamic. This may lead a patient to trust anecdotal evidence or defer choice to the surgeon, in belief that this is the optimal treatment path. They may avoid getting a second opinion, in fear of appearing disloyal.

Prospects for career advancement are also at play. For principle investigators of innovation studies, there is incentive to enroll as many patients as possible to increase the power of study and in turn have a better understanding of the true effects of treatment and publish in a higher tier journal. Thus, the physician-scientist must carefully navigate the line between patient risk and community benefit. To this point, the senior author of a clinical trial of neoadjuvant high-activity iodine-125 brachytherapy for de novo glioblastoma wrote an honest reflection of how these thoughts were present in his mind as a co-investigator [86, 92]. Berstein notes that while there is interest in professional development, the integrity and "basic virtue and trustworthiness of dedicated healthcare professionals" should prevent a clinical investigator from sacrificing the patient's well-being for the sake of individual success. Nonetheless, he underscores the usefulness of working with rigorous internal review boards (IRBs) and multidisciplinary teams who hold no or minimal vested financial or career advancement interest in the research to ensure the respect for persons, beneficence, and justice.

Context

What Is the IDEAL Approach?

Overall, it is necessary to inform patients on the nature of the outcomes and the benefits and risks of an innovate procedure, involve them in the decision-making process, and enable them to fully achieve informed consent. One way to ensure this is to impose new, more detailed regulations surrounding the requirements for consent and safe integration into practice. However, at present, there are gaps in regulation with respect to innovative procedures. Though some fear that additional policies can be unnecessarily expensive, time-consuming, and disincentivizing for patients and researchers if not carefully introduced, it is clear that additional oversight and systematic research on informed consent are required.

To address the unique qualities of surgical innovation compared to the introduction of novel pharmaceuticals, a five-stage framework was introduced through the IDEAL (Idea, Development, Exploration, Assessment, Long-term follow-up) [93, 94]. Collaboration in 2009, Stage 0 encompasses nonhuman preclinical assessments of a novel technique (animal or simulation). Stage 1 involves the first human patient for a proof-of-concept study. A prospective study by the initial surgeon in up to 30 patients comprises Stage 2a, and Stage 2b involves surgeons with no prior experience in a larger prospective study, allowing assessment of the learning curve. In Stage 3, the procedure has been optimized and should progress to a randomized controlled trial (RCT) that compares outcomes of the innovative procedure with the gold standard. Assessment of rare and long-term outcomes takes place in Stage 4.

To evaluate the feasibility of the framework and its efficacy in practice, previously, we examined two example surgical innovations: an endoscopic endonasal approach for skull base meningiomas (EEMS) and the Woven EndoBridge (WEB device) for endovascular treatment of intracranial aneurysms [95]. Our studies revealed that at the proof-of-concept stage (Stage 1), no study described ethical approval, informed consent, hospital readiness, or online registry. As studies expanded into Stage 2, there was no documentation of consent within the EEMS trials, and only 2 of 11 WEB trials reported ethical approval with patient informed consent for an innovative procedure. Overall, only 10 of the 23 clinical EEMS studies described obtaining ethical approval from an ethical commission compared to 6 out of 19 WEB device studies. We chose to spotlight this problem to note that in neurosurgery, innovations often do not follow a predefined roadmap for introduction into practice. Even though it might be challenging to adhere to the framework for every type of innovation in neurosurgery, we nevertheless believe that obtaining ethical approval for innovative procedures is feasible. Goals for future research could be to unify the informed consent procedure and outcomes reporting for emerging innovative procedures.

Recommendations

The aforementioned challenges and shortfalls in informed consent surrounding innovation in neurosurgery require us to modify current practice. Primarily, as the patients who are often candidates for unprecedented procedures may bare a disease that inhibits their capacity, we recommend that capacity screening be increased in the consent process. Ideally, this would be a quick and low-effort test, and potential tools (MacCAT-T and CCTI) and additional development may be required. Patient's ability to retain the information can be improved with more useful consent forms, media, or even group-style information sessions. Providing patient advocates and preoperative counselors, as well as an avenue for a second opinion, can help mitigate coercion.

Regarding the content of the consent, there should be as much transparency as possible. There should be full disclosure of the current data available on the method, the strengths and limitations of the technique, its preliminary outcomes, the experience of the surgeon performing the case, and alternative treatment options. As a field, introduction of these innovative methods should be done as a manner that ensures the safest translation from preclinical ideas or models to the operative room. While the IDEAL approach is comprehensive, there is evidence that the expansiveness of its recommendations are not being adhered to in practice, and additional studies to improve this may be required. Furthermore, in the process of data acquisition, it will be important for rapid dissemination of early findings to strengthen information about possible risks and benefits of new procedures and facilitate the patient information process. To ensure neurosurgeons uphold these new proposed standards of appropriateness, content, and context of informed consent, additional oversight—by peers, hospital administrators, internal review boards, or surgical societies—may be required.

Ethical Commitment and Training

Insofar, much of our discussion has centered around regulations and implementation of innovation in the neurosurgical setting. However, the individual responsibility and accountability of the physician to consistently act in accord with ethical and moral standards are paramount. The 1979 Belmont Report established basic guidelines intended to prevent ethical problems related to research three basic principles: respect for persons, beneficence, and justice [96]. These remain central to the goal of innovation, and ethical training and a moral discourse are invaluable to create a situation where investigators assume a responsibility which extends beyond obeying formal research regulations. Ethics training programs may facilitate improved ethical conduct of clinician scientists, and many research funding agencies are now requiring ethics training for scientists seeking financial support [97, 98]. We therefore believe that continued ethics training is critical for guiding practices in neurosurgical innovation.

Conclusion

Informed consent surrounding innovation in neurosurgery requires careful consideration of patient capacity and the content of discussion prior to proceeding with treatment. Ethical questions arise when a patient have cognitive impairment from disease, or when a paternalistic approach is taken at the time of information disclosure, particularly regarding surgical risk, surgeon experience, and alternative treatment options. There is inherent complexity in balancing scientific evidence, clinical expertise, and patient and family preference when pursuing innovative neurosurgical treatments. Given the degree of variability in consent practices at present, we believe modification of the field's standards for informed consent in the context of surgical innovation will lead to a higher quality of both patient care and scientific advancement.

References

1. Reardon DA, Wen PY, Wucherpfennig KW, Sampson JH. Immunomodulation for glioblastoma. Curr Opin Neurol. 2017;30(3):361–9.
2. Wiedermann JP, Joshi AS, Jamshidi A, Conchenour C, Preciado D. Utilization of a submental island flap and 3D printed model for skull base reconstruction: infantile giant craniocervicofacial teratoma. Int J Pediatr Otorhinolaryngol. 2017;92:143–5.
3. Keshavarzi S, Meltzer H, Ben-Haim S, Newman CB, Lawson JD, Levy ML, et al. Initial clinical experience with frameless optically guided stereotactic radiosurgery/radiotherapy in pediatric patients. Childs Nerv Syst. 2009;25:837–44.
4. Wong JM, Perry WR, Greenberg Y, Ho AL, Lipsitz SR, Goumnerova LC, et al. Integrating cerebrospinal fluid shunt quality checks into the World Health Organization's safe surgery checklist: a pilot study. World Neurosurg. 2016;92:491–498.e3.
5. Berg JW, Appelbaum PS, Lidz CW, Parker LS. Informed consent: legal theory and clinical practice. Systems and Psychosocial Advances Research Center Publications and Presentations, 2001;121:3–13.
6. Appelbaum PS. Clinical practice. Assessment of patients' competence to consent to treatment. N Engl J Med. 2007;357:1834–40.
7. Appelbaum PS, Grisso T. Assessing patients' capacities to consent to treatment. N Engl J Med. 1988;319:1635–8.
8. Marson D, Ingram KK. Competency to consent to treatment: a growing field of research: commentary. J Ethics Law Aging. 1996;2:59–63.
9. Okonkwo O, Griffith HR, Belue K, Lanza S, Zamrini EY, Harrell LE, et al. Medical decision-making capacity in patients with mild cognitive impairment. Neurology. 2007;69:1528–35.
10. Kerrigan S, Erridge S, Liaquat I, Graham C, Grant R. Mental incapacity in patients undergoing neuro-oncologic treatment: a cross-sectional study. Neurology. 2014;83:537–41.
11. Triebel KL, Martin RC, Nabors LB, Marson DC. Medical decision-making capacity in patients with malignant glioma. Neurology. 2009;73:2086–92.
12. Bernstein M. Neuro-oncology: under-recognized mental incapacity in brain tumour patients. Nat Rev Neurol. 2014;10:487–8.
13. Kim SY, Marson DC. Assessing decisional capacity in patients with brain tumors. Neurology. 2014;83:482–3.
14. Kirsch B, Bernstein M. Ethical challenges with awake craniotomy for tumor. Can J Neurol Sci. 2012;39:78–82.

15. Taphoorn MJ, Klein M. Cognitive deficits in adult patients with brain tumours. Lancet Neurol. 2004;3:159–68.
16. Dymek MP, Atchison P, Harrell L, Marson DC. Competency to consent to medical treatment in cognitively impaired patients with Parkinson's disease. Neurology. 2001;56:17–24.
17. Beloucif S. Informed consent for special procedures: electroconvulsive therapy and psychosurgery. Curr Opin Anaesthesiol. 2013;26:182–5.
18. Carroll D, O'Callaghan MA. Regulating, psychosurgery: ethical, social and scientific considerations. Med Law. 1984;3:193–203.
19. Hundert EM. Autonomy, informed consent, and psychosurgery. J Clin Ethics. 1994;5:264–6.
20. Mandarelli G, Moscati FM, Venturini P, Ferracuti S. Informed consent and neuromodulation techniques for psychiatric purposes: an introduction. Riv Psichiatr. 2013;48:285–92.
21. Rappaport ZH. Psychosurgery in the modern era: therapeutic and ethical aspects. Med Law. 1992;11:449–53.
22. Stagno SJ, Smith ML, Hassenbusch SJ. Reconsidering "psychosurgery": issues of informed consent and physician responsibility. J Clin Ethics. 1994;5:217–23.
23. Ramirez C, Blonski M, Belin C, Carpentier A, Taillia H. Brain metastasis: clinical and cognitive assessments. Bull Cancer. 2013;100:83–8.
24. Oishi H, Yamamoto M, Nonaka S, Arai H. Systematic review of complications for proper informed consent: (10) endovascular therapy for unruptured intracranial aneurysms. No Shinkei Geka. 2013;41:907–16.
25. Gerstenecker A, Duff K, Meneses K, Fiveash JB, Nabors LB, Triebel KL. Cognitive predictors of reasoning through treatment decisions in patients with newly diagnosed brain metastases. J Int Neuropsychol Soc. 2015;21:412–8.
26. Palmer BW, Savla GN. The association of specific neuropsychological deficits with capacity to consent to research or treatment. J Int Neuropsychol Soc. 2007;13:1047–59.
27. Cote DJ, Balak N, Brennum J, Holsgrove DT, Kitchen N, Kolenda H, et al. Ethical difficulties in the innovative surgical treatment of patients with recurrent glioblastoma multiforme. J Neurosurg. 2017;126:2045–50.
28. Kahn JP, Mastroianni AC, Sugarman J. Beyond consent: seeking justice in research, in. New York: Oxford University Press; 1998.
29. Bernat JL. Clinical ethics and the law. In: Bernat JL, editor. Ethical issues in neurology. 3rd ed. Philadelphia: Lippincott Williams & Wilkins; 2008.
30. Vyshka G, Seferi A, Myftari K, Halili V. Last call for informed consent: confused proxies in extra-emergency conditions. Indian J Med Ethics. 2014;11:252–4.
31. Adelman EE, Zahuranec DB. Surrogate decision making in neurocritical care. Continuum (Minneap Minn). 2012;18:655–8.
32. Ellamushi HE, Khan R, Kitchen ND. Consent to surgery in a high risk specialty: a prospective audit. Ann R Coll Surg Engl. 2000;82:213–6.
33. Reinert C, Kremmler L, Burock S, Bogdahn U, Wick W, Gleiter CH, et al. Quantitative and qualitative analysis of study-related patient information sheets in randomised neuro-oncology phase III-trials. Eur J Cancer. 2014;50:150–8.
34. ACOG Committee on Ethics. ACOG Committee Opinion No. 352: innovative practice: ethical guidelines. Obstet Gynecol. 2006;108:1589–95.
35. Barkun JS, Aronson JK, Feldman LS, Maddern GJ, Strasberg SM, Altman DG, et al. Evaluation and stages of surgical innovations. Lancet. 2009;374:1089–96.
36. Beard JD. Comment on: "When does the "Learning curve" of innovative interventions become questionable practice"?, P. Healey and J. Samanta, Eur J Vasc Endovasc Surg 2008;36:253-257. Eur J Vasc Endovasc Surg. 2009;37:121; author reply 121–122.
37. Biffl WL, Spain DA, Reitsma AM, Minter RM, Upperman J, Wilson M, et al. Responsible development and application of surgical innovations: a position statement of the Society of University Surgeons. J Am Coll Surg. 2008;206:1204–9.
38. Burger I, Sugarman J, Goodman SN. Ethical issues in evidence-based surgery. Surg Clin North Am. 2006;86:151–68, x.

39. Gates GA. Surgical innovation and research. Arch Otolaryngol Head Neck Surg. 2003;129:1352–3; author reply 1354.
40. Gillett G. Ethics of surgical innovation. Br J Surg. 2001;88:897–8.
41. Gillett GR. Innovative treatments: ethical requirements for evaluation. J Clin Neurosci. 1998;5:378–81.
42. Johnson J, Rogers W, Lotz M, Townley C, Meyerson D, Tomossy G. Ethical challenges of innovative surgery: a response to the IDEAL recommendations. Lancet. 2010;376:1113–5.
43. Klein E. Eloquent brain, ethical challenges: functional brain mapping in neurosurgery. Semin Ultrasound CT MR. 2015;36:291–5.
44. Lee Char SJ, Hills NK, Lo B, Kirkwood KS. Informed consent for innovative surgery: a survey of patients and surgeons. Surgery. 2013;153:473–80.
45. Lotz M. Surgical innovation as sui generis surgical research. Theor Med Bioeth. 2013;34:447–59.
46. Mastroianni AC. Liability, regulation and policy in surgical innovation: the cutting edge of research and therapy. Health Matrix Clevel. 2006;16:351–442.
47. McKneally MF. The ethics of innovation: Columbus and others try something new. J Thorac Cardiovasc Surg. 2011;141:863–6.
48. Moore FD. Ethical problems special to surgery: surgical teaching, surgical innovation, and the surgeon in managed care. Arch Surg. 2000;135:14–6.
49. Neugebauer EA, Becker M, Buess GF, Cuschieri A, Dauben HP, Fingerhut A, et al. EAES recommendations on methodology of innovation management in endoscopic surgery. Surg Endosc. 2010;24:1594–615.
50. Strasberg SM, Ludbrook PA. Who oversees innovative practice? Is there a structure that meets the monitoring needs of new techniques? J Am Coll Surg. 2003;196:938–48.
51. Sussman MD. Ethical requirements that must be met before the introduction of new procedures. Clin Orthop Relat Res. 2000;(378):15–22.
52. Wall LL, Brown D. The perils of commercially driven surgical innovation. Am J Obstet Gynecol. 2010;202(30):e31–4.
53. Qualms about innovative surgery. Lancet. 1985;1:149.
54. Healey P, Samanta J. When does the 'learning curve' of innovative interventions become questionable practice? Eur J Vasc Endovasc Surg. 2008;36:253–7.
55. Knight JL. Ethics: the dark side of surgical innovation. Innovations (Phila). 2012;7:307–13.
56. Lieberman JR, Wenger N. New technology and the orthopaedic surgeon: are you protecting your patients? Clin Orthop Relat Res. 2004;(429):338–41.
57. Tan VK, Chow PK. An approach to the ethical evaluation of innovative surgical procedures. Ann Acad Med Singap. 2011;40:26–9.
58. Wang Y, Kotsis SV, Chung KC. Applying the concepts of innovation strategies to plastic surgery. Plast Reconstr Surg. 2013;132:483–90.
59. Angelos P. The ethical challenges of surgical innovation for patient care. Lancet. 2010;376:1046–7.
60. Angelos P. Ethics and surgical innovation: challenges to the professionalism of surgeons. Int J Surg. 2013;11(Suppl 1):S2–5.
61. Miller ME, Siegler M, Angelos P. Ethical issues in surgical innovation. World J Surg. 2014;38:1638–43.
62. Shinebourne EA. Ethics of innovative cardiac surgery. Br Heart J. 1984;52:597–601.
63. Taylor PL. Overseeing innovative therapy without mistaking it for research: a function-based model based on old truths, new capacities, and lessons from stem cells. J Law Med Ethics. 2010;38:286–302.
64. Ethics Committee of American Society for Reproductive Medicine. Moving innovation to practice: a committee opinion. Fertil Steril. 2015;104:39–42.
65. Jones JW, McCullough LB, Richman BW. Ethics of surgical innovation to treat rare diseases. J Vasc Surg. 2004;39:918–9.
66. Levin AV. IOLs, innovation, and ethics in pediatric ophthalmology: let's be honest. J AAPOS. 2002;6:133–5.

67. Morgenstern L. Position statement: surgical innovations. J Am Coll Surg. 2008;207:786; author reply 786.
68. Sihvonen R, Paavola M, Malmivaara A, Itala A, Joukainen A, Nurmi H, et al. Arthroscopic partial meniscectomy versus sham surgery for a degenerative meniscal tear. N Engl J Med. 2013;369:2515–24.
69. Unger CA, Barber MD. Studying surgical innovations: challenges of the randomized controlled trial. J Minim Invasive Gynecol. 2015;22:573–82.
70. Honeybul S, O'Hanlon S, Ho KM, Gillett G. The influence of objective prognostic information on the likelihood of informed consent for decompressive craniectomy: a study of Australian anaesthetists. Anaesth Intensive Care. 2011;39:659–65.
71. Wells RE. To tell the truth, the whole truth, may do patients harm: the problem of the nocebo effect for informed consent. Am J Bioeth. 2012;12:22–9.
72. Amanzio M, Corazzini LL, Vase L, Benedetti F. A systematic review of adverse events in placebo groups of anti-migraine clinical trials. Pain. 2009;146:261–9.
73. Rief W, Avorn J, Barsky AJ. Medication-attributed adverse effects in placebo groups: implications for assessment of adverse effects. Arch Intern Med. 2006;166:155–60.
74. Krupp W, Spanehl O, Laubach W, Seifert V. Informed consent in neurosurgery: patients' recall of preoperative discussion. Acta Neurochir. 2000;142:233–8; discussion 238–239.
75. Daugherty CK. Impact of therapeutic research on informed consent and the ethics of clinical trials: a medical oncology perspective. J Clin Oncol. 1999;17:1601–17.
76. Rozmovits L, Khu KJ, Osman S, Gentili F, Guha A, Bernstein M. Information gaps for patients requiring craniotomy for benign brain lesion: a qualitative study. J Neuro-Oncol. 2010;96:241–7.
77. Wozniak RH. Classics in the history of psychology—introduction to Ebbinghaus. Bristol: Thoemmes Press; 1999. p. 1885/1913.
78. Harel R, Pfeffer R, Levin D, Shekel E, Epstein D, Tsvang L, et al. Spine radiosurgery: lessons learned from the first 100 treatment sessions. Neurosurg Focus. 2017;42:E3.
79. van der Linden W. Pitfalls in randomized surgical trials. Surgery. 1980;87:258–62.
80. Silva PS, Pereira P, Monteiro P, Silva PA, Vaz R. Learning curve and complications of minimally invasive transforaminal lumbar interbody fusion. Neurosurg Focus. 2013;35:E7.
81. Chi F, Wang Y, Lin Y, Ge J, Qiu Y, Guo L. A learning curve of endoscopic transsphenoidal surgery for pituitary adenoma. J Craniofac Surg. 2013;24:2064–7.
82. Shikary T, Andaluz N, Meinzen-Derr J, Edwards C, Theodosopoulos P, Zimmer LA. Operative learning curve after transition to endoscopic transsphenoidal pituitary surgery. World Neurosurg. 2017;102:608–12.
83. Bilotta F, Titi L, Lanni F, Stazi E, Rosa G. Training anesthesiology residents in providing anesthesia for awake craniotomy: learning curves and estimate of needed case load. J Clin Anesth. 2013;25:359–66.
84. Physician's experience as an element of informed consent - Johnson v. Kokemoor, in Wisconsin SCo (ed). 545 N.W.2d 495, 199 Wis. 2d 615 (Wi): VersusLaw Inc.; 1996.
85. Banja JD. Disclosure of experience as a risk factor in informed consent for neurosurgery: the case of Johnson v. Kokemoor. Virtual Mentor. 2015;17:69–73.
86. Bernstein M. Conflict of interest: it is ethical for an investigator to also be the primary caregiver in a clinical trial. J Neuro-Oncol. 2003;63:107–8.
87. Backous DD, Pham HT. Guiding patients through the choices for treating vestibular schwannomas: balancing options and ensuring informed consent. Otolaryngol Clin N Am. 2007;40:521–40, viii–ix.
88. Robertson FC, Logsdon JL, Dasenbrock HH, Yan SC, Raftery SM, Smith TR, et al. Transitional care services: a quality and safety process improvement program in neurosurgery. J Neurosurg. 2018;128(5):1570–7.
89. Ford PJ. Vulnerable brains: research ethics and neurosurgical patients. J Law Med Ethics. 2009;37:73–82.
90. McKneally MF. Ethical problems in surgery: innovation leading to unforeseen complications. World J Surg. 1999;23:786–8.

91. Bracken-Roche D, Bell E, Karpowicz L, Racine E. Disclosure, consent, and the exercise of patient autonomy in surgical innovation: a systematic content analysis of the conceptual literature. Account Res. 2014;21:331–52.
92. Laperriere NJ, Leung PM, McKenzie S, Milosevic M, Wong S, Glen J, et al. Randomized study of brachytherapy in the initial management of patients with malignant astrocytoma. Int J Radiat Oncol Biol Phys. 1998;41:1005–11.
93. Hirst A, Agha RA, Rosin D, McCulloch P. How can we improve surgical research and innovation?: the IDEAL framework for action. Int J Surg. 2013;11:1038–42.
94. McCulloch P, Altman DG, Campbell WB, Flum DR, Glasziou P, Marshall JC, et al. No surgical innovation without evaluation: the IDEAL recommendations. Lancet. 2009;374:1105–12.
95. Muskens I, Diederen SJH, Senders JT, Zamanipoor Najafabadi AH, van Furth WR, May AM, Smith TR, Bredenoord AL, Broekman MLD. Innovation in neurosurgery: less than IDEAL? A systematic review. Acta Neurochir. 2017;159(10):1957–66.
96. Department of Health, Education, and Welfare, National Commission for the Protection of Human Subjects of Biomedical and Behavioral Research. The Belmont Report. Ethical principles and guidelines for the protection of human subjects of research. J Am Coll Dent. 2014;81:4–13.
97. Dalton R. NIH cash tied to compulsory training in good behaviour. Nature. 2000;408:629.
98. Mumford MD, Steele L, Watts LL. Evaluating Ethics Education Programs: a multilevel approach. Ethics Behav. 2015;25:37–60.

Ethical Challenges of Current Oversight and Regulation of Novel Medical Devices in Neurosurgery

3

Ivo S. Muskens, Saksham Gupta, Alexander F. C. Hulsbergen, Wouter A. Moojen, and Marike L. D. Broekman

Introduction

Since its inception, outcomes of neurosurgical procedures have been tremendously improved by innovation of medical devices. Medical devices are meant to prevent, diagnose, or treat disease and may be instruments, implants, or even mechanical

This chapter is based on: Muskens IS, Gupta S, Hulsbergen A, Moojen WA, Broekman ML. Introduction of Novel Medical Devices in Surgery: Ethical Challenges of Current Oversight and Regulation. J Am Coll Surg. 2017 Aug 4.

I. S. Muskens
Department of Neurosurgery, Haaglanden Medical Center, The Hague, The Netherlands

Department of Neurosurgery, Leiden University Medical Center, Leiden, The Netherlands

Department of Neurosurgery, Computational Neurosciences Outcomes Center (CNOC), Brigham and Women's Hospital, Harvard Medical School, Boston, MA, USA

S. Gupta
Department of Neurosurgery, Computational Neurosciences Outcomes Center (CNOC), Brigham and Women's Hospital, Harvard Medical School, Boston, MA, USA

W. A. Moojen
Department of Neurosurgery, Haaglanden Medical Center, The Hague, The Netherlands

Department of Neurosurgery, Leiden University Medical Center, Leiden, The Netherlands

Department of Neurosurgery, Haga Teaching Hospital, The Hague, The Netherlands

M. L. D. Broekman (⊠) · A. F. C. Hulsbergen
Department of Neurosurgery, Haaglanden Medical Center, The Hague, The Netherlands

Department of Neurosurgery, Leiden University Medical Center, Leiden, The Netherlands

Department of Neurosurgery, Computational Neurosciences Outcomes Center (CNOC), Brigham and Women's Hospital, Harvard Medical School, Boston, MA, USA
e-mail: m.broekman@haaglandenmc.nl; m.l.d.broekman@lumc.nl

© Springer Nature Switzerland AG 2019
M. L. D. Broekman (ed.), *Ethics of Innovation in Neurosurgery*,
https://doi.org/10.1007/978-3-030-05502-8_3

agents [1]. Unfortunately, novel devices have not always turned out to be an improvement over existing treatments, and some have even led to severe complications in patients. For instance, in neurosurgery, interspinous devices (IDs) were approved and used for several years before studies evaluating safety showed major unforeseen side effects [2–4].

Different countries have created various forms of regulation to ensure safety of medical devices and protect patients during investigation of novel devices. Both in the United States of America (USA) and the European Economic Area (EEA: the European Union (EU), Switzerland, Lichtenstein, Norway, and Iceland), several agencies are responsible for approval of medical devices before they are allowed on the market. However, gaps and differences in respective regulations leave room for various ethical challenges that relate to risk-benefit ratio, informed consent, scientific validity, societal value, and justice. In this chapter, the current legislative landscape surrounding medical device introduction is evaluated for the USA and EEA, respectively, and arising ethical challenges are evaluated and addressed.

Overview of Current Regulation

Both the Food and Drug Administration (FDA) and Conformité Européenne (CE) are responsible government agencies for medical device evaluation and approval in the USA and EEA, respectively. For clinical application of medical devices in the USA and EEA, FDA approval and CE marking are required, respectively (Table 3.1).

FDA

If a medical device company wants to apply for FDA approval, it first has to register with the FDA and have its device classified [5, 6]. The FDA states:

Table 3.1 Overview of the approval process for CE marking and FDA approval

Approval process		CE marking	FDA approval
Device classification	Class I/II	National competent authorities approve the device based on safety and efficacy date	510(k)
	Class III	Approval by Notified Bodies based on safety and efficacy studies	PMA study
Exemptions from approval		Devices deemed by national competent authorities to have low risk	IDE, 510(k) exemption, humanitarian exception
Post-market evaluation		EUDAMED	MAUDE database, MEDSUN device, "522 study"

CE Conformité Européenne, *FDA* Food and Drug Administration, *PMA* premarket approval, *MAUDE* Manufacturer and User Facility Device Experience, *IDE* Investigational Device Exemption, *EUDAMED* European Database on Medical Devices

"Device classification depends on the intended use of the device and also upon indications for use." … "In addition, classification is risk based, that is, the risk the device poses to the patient and/or the user is a major factor in the class it is assigned." [7]

Furthermore, the Product Code Classification Database encompasses various examples of devices and their respective classification, but does not provide a strict guideline as such [8, 9]. According to this database, surgical gloves and instruments are examples of Class I devices. Surgical meshes, EEG electrodes, aneurysm clips, and endovascular embolization materials are examples of Class II devices. Invasive devices that generate or modulate biological signals, such as spinal and deep brain stimulators, form Class III devices [10].

If a medical device company wishes to have its Class I or II device approved, it has to provide premarket notification (510(k)) [11]. This notification must provide evidence that the medical device is safe and effective compared to a "substantially equivalent" medical device [11]. However, the majority of Class I and II devices are exempt from the 510(k) process due to expected similar performance of the device [8, 12]. For Class III devices, the FDA requires premarket approval studies (PMA) [5, 13]. Once the device is approved, modifications to the device may be applied through the PMA supplement pathway and are not uncommon and rarely supported by data from additional trials [14–17].

The FDA has the right to demand a post-market surveillance studies ("522 studies") that are aimed at identifying rare adverse events and long-term complications [18–20]. The FDA also has the right to remove a medical device from the market for safety reasons, but not for efficacy reasons [17]. The FDA also receives safety and efficacy data for approved devices through the Manufacturer and User Facility Device Experience (MAUDE) registry to which physicians, manufacturers, and patients can report complications [21]. Additionally, the FDA works together with 280 hospitals in the online "Medical Product Safety Network" (MedSun) adverse event program [22, 23].

In some circumstances FDA approval is not required for clinical use of the medical device. For instance, emergent use, compassionate use, or evaluation of a medical device in a trial is allowed through the Investigational Device Exemption (IDE) [9, 24–26]. Furthermore, rare diseases may be treated through "humanitarian device exception" without formal approval [22].

CE Marking

In the EEA and Turkey, manufacturers are required to acquire CE marking for their medical device before it is allowed onto the market [27–30]. An authorized representative within the EEA is required for manufacturers outside the EEA [31]. Devices are classified into three classes based on invasiveness, reusability, potential use as an implant, the use of a power source, and the use near a critical anatomical location [32, 33]. For low-risk devices, several EU-appointed national competent authorities, such as the Medicines and Healthcare Products Regulatory Agency

(MHRA) in the UK, may grant the CE mark [22]. Private EU-authorized, third-party Notified Bodies are responsible for the approval of higher-risk devices and the review of safety and efficacy of the device [22, 34, 35]. Unlike Class III devices in the USA that require PMA studies, a trial is not necessarily required for CE marking for high-risk devices in the EEA [22, 27]. For post-market surveillance, the European Database on Medical Devices (EUDAMED) provides national competent authorities with data on safety medical devices [36].

Some have suggested that the CE-marking approval process is inconsistent due to the different Notified Bodies and the fact that approval from only one authority is required [22, 37]. Medical device manufacturers may choose to approach the Notified Bodies with the least stringent approval to save costs. To prove this point, a group of Dutch journalists approached three different Notified Bodies with a tangerine net that was to be used for prolapse repair and got a reported likelihood of approval greater than 90% [38].

Off-Label Use

Off-label use of medical devices is allowed both in the USA and Europe as long as the aim is to "practice medicine" [26, 39]. Off-label use of medical pharmaceuticals was recently independently associated with higher rates of adverse events compared to on-label use, which may also be the case for medical devices [40]. Risks associated with off-label use of medical devices may even be greater because of different tissue properties in different pathologies and different anatomical characteristics. For instance, off-label use of rhBMP in cervical spine resulted in osteolysis, hematomas, and dysphagia and even lead the FDA to issue a formal Public Health Notification Warning [41–43].

Ethical Challenges

The legislative landscape outlined above leaves room for several ethical challenges with regard to risk-benefit ratio, informed consent, scientific validity, societal value, and justice.

Risk-Benefit Ratio

For every treatment or procedure, associated benefits should always outweigh the risks, and medical devices are of course no exception. However, risks and benefits may be hard to anticipate if the provided evidence is based on preliminary studies or on experience with other pathology. The knowledge of the risk-benefit ratio may be further obfuscated by a possible lack of standardization of clinical studies, varying quality of trials, and ineffective post-market surveillance [13, 20].

Medical devices may be used in situation with poorly defined risk-benefit ratios due to several legislative loopholes. For instance, 510(k) exemption allows approval for Class I and II devices without any clinical evaluation [8, 12]. In the EEA, the situation may even be more challenging as even high-risk devices may not even require a clinical study [22, 27, 32]. Furthermore, inconsistent approval and variation among Notified Bodies are suggested to further limit efficacy and safety data [22, 27, 32]. Off-label use inherently is not supported by any rigorous evaluation and may therefore also compromise patient outcomes as efficacy and safety are unknown.

Informed Consent

Adequate informed consent is essential for patients to be able to decide whether to proceed with the proposed treatment and requires explanation of potential risks and benefits. The patients' autonomy may, therefore, not be respected if the poorly defined risk-benefit ratios obfuscate the informed consent process. For instance, weak data produced by low-quality trails will limit the patients' ability to make adequate decisions [13, 20, 22, 27, 32].

This is further complicated by inaccessibility and incomprehensibility of many databases on adverse events [23, 36, 44]. For off-label use, it is not required that a patient is informed that the device is not being used for its intended purpose. Finally, the informed consent process is severely compromised in the case of Class III medical devices in the EEA as no clinical evaluation is required for approval [22].

Scientific Validity

A scientifically valid treatment is one of the main pillars for the trust patients put in their neurosurgeon. Evaluation of medical devices may range from preliminary preclinical study to a randomized controlled trial (RCT), the pinnacle of trial design. However, it is unfeasible to require an RCT for every medical device to make sure innovation is not stifled, as an RCT is costly for private device manufacturers and is time-consuming. This does not mean that medical devices do not require scientifically valid evidence. A 510(k) exemption and the lack of requirement for trials for a CE marking do not ensure evidence-based practice, which may increase risks for patients and lead to maleficence [8, 12, 22, 27, 32].

To introduce medical devices in an ethical fashion, the Idea, Development, Exploration, Assessment, Long-term Follow-up (IDEAL) consortium of surgeons, statisticians, and epidemiologists has proposed the IDEAL-Device (IDEAL-D) framework [45]. The IDEAL-D framework suggests that a randomized comparison should be performed with the current standard of care as reference [45]. However, the FDA does not require comparison with the gold standard, which led to the comparison of IDs with other devices instead of the gold standard lumbar decompression [2, 17, 46]. PMA studies have also been suggested to vary greatly in terms of quality [13]. This is further complicated by off-label use of devices, which is only supported by expert opinion.

Societal Value

The benefit of a medical device must outweigh its associated costs for society to be ethical. Industry reports suggest that a 510(k) approval costs $34 million and PMA studies cost $94 million, which may be regarded as costly [47]. No infrastructure is in place to evaluate societal value of medical devices. The FDA is even prohibited from cost-effectiveness analysis in the approval process [17]. Similarly, cost-effect analysis is further complicated as 50% of side effects in drugs are only discovered after FDA approval, which may be similar in medical devices [17]. As a result, healthcare costs associated with medical devices may continue to increase for society without any appreciable benefit.

Justice

Availability and associated risks of medical devices have to be shared equally among potential patients to do justice. The lower costs associated with the approval process in Europe result in many manufacturers opting to introduce their device there first [48]. On the one hand, European patients may have earlier access as compared to the USA but may also be exposed to increased risk associated with relatively unproven devices [48].

Potentials for Improvement of Oversight and Regulation for All Involved Parties

The device manufacturer, the regulation authority, the surgeon, and the patient all can improve the legislative landscape for the introduction of medical devices because of respective responsibilities (Table 3.2). All parties should share the same values, patient safety, patient autonomy, surgeon support, the facilitation of evidence-based practice, and a climate of continuous innovation.

Legislator and Oversight Bodies

The legislator and oversight bodies could alter current medical device legislation to remove lapses and try to move the incentives for manufactures from financial gain to device safety and efficacy. By requiring Level 2 evidence prior to approval and provide funding for Level 1 evidence, the value of FDA approval and CE marking can be improved. This may be achieved by mandating adherence to the IDEAL-D framework for high-risk devices [45]. This mandate should also limit the disparity

Table 3.2 Responsibilities for all parties involved to improve regulation and oversight for the use of medical devices

Involved party	Possible means of improvement
Legislator and oversight bodies	– Create financial incentive for manufacturers to produce more safety and efficacy data – Demand for clinical data showing efficacy and safety for devices that are currently approved through exemption – Improve requirements for Class III device evaluation in Europe – Create alternative funding for device development (e.g., NIH) – Introduce oversight for off-label use of medical devices (e.g., a mandatory registry)
Medical device manufacturer	– Change business model to produce more safety and efficacy data – Introduce medical devices simultaneously in the USA and Europe – Introduction of industry guidelines for ethical introduction of medical devices
Surgeon	– Present conflicts of interest to the patient through a mandatory statement – Participate in registries and surgical innovation Committees for off-label use of medical devices
Patient	– Active participation in device development – Allow sharing of their data to aid evaluation of medical devices

in regulation between the USA and Europe and end the current practice of earlier introduction of devices in Europe.

Alternatively, government bodies could provide grants to stimulate decisions by manufacturers that are aimed at long-term benefit instead of financially driven decisions with short-term aims [45]. Introducing a cap on the amount of funding by private parties could improve the current situation, as private funding is known to influence outcomes of pharmaceutical trials [49, 50].

The practice of off-label use of medical devices may also be improved by introducing regulation that regards medical devices differently from pharmaceuticals. One solution could be a requirement that every off-label procedure is registered with an oversight body in a registry to ensure the possibility to evaluate outcomes. This would still respect the judgment of the surgeon and ensure patient safety.

The evaluation of long-term outcomes and identification of rare adverse events may be ensured through the introduction of a centralized registry such as the "National Evaluation System for Health Technology," which has been shown to be beneficial in identifying adverse events in a cardiac medical device [51–53]. Together with post-market surveillance, this could limit the time, and unsafe device is allowed onto the market.

Medical Device Manufacturer

Every medical device manufacturer has the responsibility to provide safe and effective medical device. This responsibility may be compromised by financial incentives that encourage manufacturers to require reimbursement for their device before anything else [54]. Self-regulation in the form of guidelines by the Medical Device Manufacturers Association could impose ethical standards for the introduction of devices, comparable to the food industry [55]. Transparency may be improved by the medical device industry by collaboration with oversight bodies, surgeons, and patients and through generating and providing extensive safety and efficacy data, as seen in the aviation industry [56].

Neurosurgeon

Neurosurgeons hold a key position in the introduction of medical devices. Through both conscious and creative decisions, they continuously innovate while constantly balancing risks and benefits for patients. However, this judgment on risk-benefit ratio may be clouded by financial and professional conflicts of interest (COIs). This may be particularly the case in Europe, as currently no uniform legislation exists [45, 57, 58]. Transparency toward patients may be improved by including all financial gains for surgeons from private parties in a publically accessible database [45, 57, 58]. In the USA, the Sunshine Act requires that all industry payments to neurosurgeons are registered in a transparent database. This database showed that neurosurgeons received $30,718.02 on average from companies in 2014 [59, 60]. The amount of money a neurosurgeon receives from the industry could be limited by a cap to limit COI. Neurosurgeons could also be made to register use and outcomes of medical devices for which COIs exist. A statement toward patients that contains the manufacturer, the amount of compensation, and alternative treatment options could improve transparency. The informed consent procedure could be further improved by description of all available evidence of safety and efficacy for the device that is to be used [22, 27].

Neurosurgeons could also actively participate in registries for off-label use of medical devices to ensure scientific validity and the possibility to study outcomes. The neurosurgeon could be made responsible for the evaluation of surgical learning curve, long-term functional outcomes, and device-specific adverse events. Surgical innovation committees (SIC) could also provide a forum for surgeons to discuss and evaluate device-related innovation [61]. Appropriate oversight over the medical device innovation could be one of the responsibilities of the SIC.

Patients

Input from patients is essential to introduce medical devices in an ethically sound manner, as patients receive treatment. Patients are also partially responsible for the quality of care for future patients as their treatment is the result of risks taken by previous patients [62, 63]. Patient organizations could work together with manufacturers and legislators in setting priorities for medical devices. Patients are the best judges when it comes to defining acceptable risks and meaningful benefits. Safety monitoring could also be improved by patients allowing their electronic healthcare records to be shared into a centralized database [62]. Despite this, patients should never be responsible for outcomes, evaluation, or safety monitoring as oversight bodies should do this on their behalf.

Conclusion

Many ethical challenges arise from the current oversight and regulation for the introduction of medical devices in neurosurgery. Innovation and patient safety need to be balanced by appropriate oversight. However, current regulation leaves multiple gaps that result in ethical challenges regarding risk-benefit ratio, informed consent, scientific validity, societal value, and justice which are described in this chapter for the neurosurgeon to consider. The situation may be improved through changes of current oversight mechanisms and legislation and by increasing awareness about the shared responsibilities of all involved parties. This could ensure ethical introduction of medical devices and improve the quality of neurosurgical care, the ultimate goal.

References

1. US Food and Drug Administration. What is a medical device? [cited 2015]. https://www.fda.gov/AboutFDA/Transparency/Basics/ucm211822.htm.
2. Moojen WA, Arts MP, Jacobs WC, van Zwet EW, van den Akker-van Marle ME, Koes BW, et al. Interspinous process device versus standard conventional surgical decompression for lumbar spinal stenosis: randomised controlled trial. Br J Sports Med. 2015;49(2):135.
3. Moojen WA, Arts MP, Jacobs WC, van Zwet EW, van den Akker-van Marle ME, Koes BW, et al. Interspinous process device versus standard conventional surgical decompression for lumbar spinal stenosis: randomized controlled trial. BMJ. 2013;347:f6415.
4. Griffin WL, Nanson CJ, Springer BD, Davies MA, Fehring TK. Reduced articular surface of one-piece cups: a cause of runaway wear and early failure. Clin Orthop Relat Res. 2010;468(9):2328–32.
5. US Food and Drug Administration. CFR - Code of Federal Regulations Title 21. 2016. http://www.accessdata.fda.gov/scripts/cdrh/cfdocs/cfcfr/CFRSearch.cfm?fr=807.20.
6. Chohan MO, Levin AM, Singh R, Zhou Z, Green CL, Kazam JJ, et al. Three-dimensional volumetric measurements in defining endoscope-guided giant adenoma surgery outcomes. Pituitary. 2016;19(3):311–21.
7. US Food and Drug Administration. Classify your medical device. 2014. https://www.fda.gov/medicaldevices/deviceregulationandguidance/overview/classifyyourdevice/.
8. US Food and Drug Administration. The 510(k) Program: evaluating substantial equivalence in premarket notifications [510(k)] Guidance for Industry and Food and Drug Administration staff. 2014. http://www.fda.gov/downloads/MedicalDevices/DeviceRegulationandGuidance/GuidanceDocuments/UCM284443.pdf.
9. US Food and Drug Administration. Device Classification Panels 2017. https://www.fda.gov/MedicalDevices/DeviceRegulationandGuidance/Overview/ClassifyYourDevice/ucm051530.htm.
10. US Food and Drug Administration. Title 21—Food and drugs chapter I—Food and Drug Adminitration Department of Health and Human Survices subchapter H 2016. http://www.accessdata.fda.gov/scripts/cdrh/cfdocs/cfcfr/CFRSearch.cfm?CFRPart=882.

11. US Food and Drug Administration. Code of Federal Regulations Title 21 Part 807: establishment registration and device listing for manufacutrers and initial importors of devices subpart E—premarket notification procedures 2016. http://www.accessdata.fda.gov/scripts/cdrh/cfdocs/cfcfr/CFRSearch.cfm?CFRPart=807&showFR=1&subpartNode=21:8.0.1.1.5.5.

12. US Food and Drug Administration. Medical device exemptions 510(k) and GMP requirements 2016. https://www.accessdata.fda.gov/scripts/cdrh/cfdocs/cfpcd/315.cfm.

13. Golish SR, Reed ML. Spinal devices in the United States-Investigational Device Exemption Trials and premarket approval of class III devices. Spine J. 2017;17(1):150–7.

14. Samuel AM, Rathi VK, Grauer JN, Ross JS. How do orthopaedic devices change after their initial FDA premarket approval? Clin Orthop Relat Res. 2016;474(4):1053–68.

15. Rome BN, Kramer DB, Kesselheim AS. FDA approval of cardiac implantable electronic devices via original and supplement premarket approval pathways, 1979-2012. JAMA. 2014;311(4):385–91.

16. US Food and Drug Administration. PMA supplements and amendments 2002. https://www.fda.gov/MedicalDevices/DeviceRegulationandGuidance/HowtoMarketYourDevice/PremarketSubmissions/PremarketApprovalPMA/ucm050467.htm.

17. Carragee EJ, Deyo RA, Kovacs FM, Peul WC, Lurie JD, Urrutia G, et al. Clinical research: is the spine field a mine field? Spine (Phila Pa 1976). 2009;34(5):423–30.

18. US Food and Drug Administration. Current postmarket surveillance efforts 2014 [April 10, 2017]. http://www.fda.gov/MedicalDevices/Safety/CDRHPostmarketSurveillance/ucm348738.htm.

19. Rome BN, Kramer DB, Kesselheim AS. Approval of high-risk medical devices in the US: implications for clinical cardiology. Curr Cardiol Rep. 2014;16(6):489.

20. Reynolds IS, Rising JP, Coukell AJ, Paulson KH, Redberg RF. Assessing the safety and effectiveness of devices after US Food and Drug Administration approval: FDA-mandated postapproval studies. JAMA Intern Med. 2014;174(11):1773–9.

21. US Food and Drug Administration. How the FDA will create an integrated national postmarket surveillance system 2014 [April 10, 2017]. http://www.fda.gov/MedicalDevices/Safety/CDRHPostmarketSurveillance/ucm348751.htm.

22. Kramer DB, Xu S, Kesselheim AS. Regulation of medical devices in the United States and European Union. N Engl J Med. 2012;366(9):848–55.

23. US Food and Drug Administration. MedSun: Medical Product Safety Network 2016. https://www.fda.gov/MedicalDevices/Safety/MedSunMedicalProductSafetyNetwork/.

24. Blumenthal S, McAfee PC, Guyer RD, Hochschuler SH, Geisler FH, Holt RT, et al. A prospective, randomized, multicenter Food and Drug Administration Investigational Device Exemptions Study of lumbar total disc replacement with the CHARITE artificial disc versus lumbar fusion: part I: evaluation of clinical outcomes. Spine (Phila Pa 1976). 2005;30(14):1565–75; discussion E387–91

25. US Food and Drug Administration. Code of Federal Regulations; Title 21—Food and drug, Chapter I—Food and Drug Administration Department of Health and Human Servises subchapter H—Medial devices part 812 investigational device exemptions 2016. http://www.accessdata.fda.gov/scripts/cdrh/cfdocs/cfcfr/CFRSearch.cfm?CFRPart=812&showFR=1.

26. US Food and Drug Administration. Guidance on IDE policies and procedures 2015 [April 10, 2017]. http://www.fda.gov/MedicalDevices/DeviceRegulationandGuidance/GuidanceDocuments/ucm080202.htm.

27. European Commission. Clinical evaluation: a guide for manufacturers and notified bodies (MEDDEV. 2.7.1 Rev.3) 2009 [April 10, 2017]. http://ec.europa.eu/consumers/sectors/medical-devices/files/meddev/2_7_1rev_3_en.pdf.

28. Fraser AG, Daubert JC, Van de Werf F, Estes NA III, Smith SC Jr, Krucoff MW, et al. Clinical evaluation of cardiovascular devices: principles, problems, and proposals for European regulatory reform. Report of a policy conference of the European Society of Cardiology. Eur Heart J. 2011;32(13):1673–86.

29. Schweizerische Eidgenossenschaft. Abkommen zwischen der Schweizerischen Eidgenossenschaft und der Europäischen Gemeinschaft über die gegenseitige Anerkennung

von Konformitätsbewertungen 1999 [April 10, 2017]. https://www.admin.ch/opc/de/classified-compilation/19994644/index.html.

30. European Commission. Interpretation of the customs union agreement with Turkey in the field of medical devices 2010 [updated April 10, 2017]. http://ec.europa.eu/DocsRoom/documents/10270/attachments/1/translations/en/renditions/native.

31. European Commission. Interpretative document of the commision's placing on the market of medical device 2010 [April 10, 2017]. http://ec.europa.eu/consumers/sectors/medical-devices/files/guide-stds-directives/placing_on_the_market_en.pdf.

32. Hulstaert F, Neyt M, Vinck I, Stordeur S, Huic M, Sauerland S, et al. Pre-market clinical evaluations of innovative high-risk medical devices in Europe. Int J Technol Assess Health Care. 2012;28(3):278–84.

33. European Commission. MEDICAL DEVICES: Guidance document - classification of medical devices 2010. http://ec.europa.eu/consumers/sectors/medical-devices/files/meddev/2_4_1_rev_9_classification_en.pdf.

34. Kramer DB, Xu S, Kesselheim AS. How does medical device regulation perform in the United States and the European union? A systematic review. PLoS Med. 2012;9(7):e1001276.

35. Heinemann L, Freckmann G, Koschinsky T. Considerations for an institution for evaluation of diabetes technology devices to improve their quality in the European Union. J Diabetes Sci Technol. 2013;7(2):542–7.

36. The Eudamed Working Group. Evaluation of the "EUropean DAtabank on MEdical Devices" 2012 [April 10, 2017]. http://ec.europa.eu/growth/sectors/medical-devices/market-surveillance_nl.

37. Cohen D, Billingsley M. Europeans are left to their own devices. BMJ. 2011;342:d2748.

38. RADAR. Mandarijnennetje als implantaat: goedkeuring implantaten is een farce 2014 [April 10, 2017]. http://radar.avrotros.nl/nieuws/detail/mandarijnennetje-als-implantaat-goedkeuring-implantaten-is-een-farce/.

39. US Food and Drug Administration. "Off-Label" and investigational use of marketed drugs, biologics, and medical devices - Information Sheet 2016. http://www.fda.gov/RegulatoryInformation/Guidances/ucm126486.htm.

40. Eguale T, Buckeridge DL, Verma A, Winslade NE, Benedetti A, Hanley JA, et al. Association of off-label drug use and adverse drug events in an adult population. JAMA Intern Med. 2016;176(1):55–63.

41. Epstein NE. Complications due to the use of BMP/INFUSE in spine surgery: the evidence continues to mount. Surg Neurol Int. 2013;4(Suppl 5):S343–52.

42. Schnurman Z, Smith ML, Kondziolka D. Off-label innovation: characterization through a case study of rhBMP-2 for spinal fusion. J Neurosurg Spine. 2016;25(3):406–14.

43. US Food and Drug Administration. FDA public health notification: life-threatening complications associated with recombinant human bone morphogenetic protein in cervical spine fusion 2008 [April 10, 2017]. http://www.fda.gov/MedicalDevices/Safety/AlertsandNotices/PublicHealthNotifications/ucm062000.htm.

44. Coelho DH, Tampio AJ. The utility of the MAUDE database for osseointegrated auditory implants. Ann Otol Rhinol Laryngol. 2017;126(1):61–6.

45. Sedrakyan A, Campbell B, Merino JG, Kuntz R, Hirst A, McCulloch P. IDEAL-D: a rational framework for evaluating and regulating the use of medical devices. BMJ. 2016;353:i2372.

46. Moojen WA, Bredenoord AL, Viergever RF, Peul WC. Scientific evaluation of spinal implants: an ethical necessity. Spine (Phila Pa 1976). 2014;39(26):2115–8.

47. Makower J, Meed A, Denend L. FDA impact on U.S. medical technology innovation a survey of over 200 Medical Technology Companies; 2010.

48. Hwang TJ, Sokolov E, Franklin JM, Kesselheim AS. Comparison of rates of safety issues and reporting of trial outcomes for medical devices approved in the European Union and United States: cohort study. BMJ. 2016;353:i3323.

49. Lexchin J, Bero LA, Djulbegovic B, Clark O. Pharmaceutical industry sponsorship and research outcome and quality: systematic review. BMJ. 2003;326(7400):1167–70.

50. Lundh A, Sismondo S, Lexchin J, Busuioc OA, Bero L. Industry sponsorship and research outcome. Cochrane Database Syst Rev. 2012;(12):MR000033.

51. Shuren J, Califf RM. Need for a National Evaluation System for Health Technology. JAMA. 2016;316(11):1153–4.
52. US Food and Drug Administration. National medical device postmarket surveillance plan 2012 [April 10, 2017]. http://www.fda.gov.proxy.library.uu.nl/AboutFDA/CentersOffices/OfficeofMedicalProductsandTobacco/CDRH/CDRHReports/ucm301912.htm.
53. Resnic FS, Majithia A, Marinac-Dabic D, Robbins S, Ssemaganda H, Hewitt K, et al. Registry-based prospective, active surveillance of medical-device safety. N Engl J Med. 2017;376:526–35.
54. Nelsen LL, Bierer BE. Biomedical innovation in academic institutions: mitigating conflict of interest. Sci Transl Med. 2011;3(100):100cm26.
55. Sharma LL, Teret SP, Brownell KD. The food industry and self-regulation: standards to promote success and to avoid public health failures. Am J Public Health. 2010;100(2):240–6.
56. Lewis GH, Vaithianathan R, Hockey PM, Hirst G, Bagian JP. Counterheroism, common knowledge, and ergonomics: concepts from aviation that could improve patient safety. Milbank Q. 2011;89(1):4–38.
57. Perlis RH, Perlis CS. Physician payments from industry are associated with greater medicare part D prescribing costs. PLoS One. 2016;11(5):e0155474.
58. Arie S. The device industry and payments to doctors. BMJ. 2015;351:h6182.
59. Institute of Medicine Committee on Conflict of Interest in Medical Research Eeuctaion, and Practice. The National Academies Collection: reports funded by National Institutes of Health. In: Lo B, Field MJ, editors. Conflict of interest in medical research, education, and practice. Washington, DC: National Academies Press (US) National Academy of Sciences; 2009.
60. Babu MA, Heary RF, Nahed BV. Does the open payments database provide sunshine on neurosurgery? Neurosurgery. 2016;79(6):933–8.
61. Biffl WL, Spain DA, Reitsma AM, Minter RM, Upperman J, Wilson M, et al. Responsible development and application of surgical innovations: a position statement of the Society of University Surgeons. J Am Coll Surg. 2008;206(6):1204–9.
62. Faden RR, Kass NE, Goodman SN, Pronovost P, Tunis S, Beauchamp TL. An ethics framework for a learning health care system: a departure from traditional research ethics and clinical ethics. Hastings Cent Rep. 2013;Spec No:S16–27.
63. Faden RR, Beauchamp TL, Kass NE. Informed consent, comparative effectiveness, and learning health care. N Engl J Med. 2014;370(8):766–8.

Saksham Gupta, Ivo S. Muskens, Luis Bradley Fandino,
Alexander F. C. Hulsbergen, and Marike L. D. Broekman

Introduction

Innovation within the operating room has advanced neurosurgical outcomes. This type of innovation may range from minor adjustments to gold standard protocols to novel procedures that address a patient's unique anatomy. While formal research is regulated closely by institutional review boards (IRB's), surgical innovations often fall outside this realm for multiple reasons. The initial barriers remain defining surgical innovation and distinguish it from research. The distinction between surgical innovation within clinical practice and research can be hard to make. According to

This chapter is based on: Gupta S, Muskens IS, Fandino LB, Hulsbergen AFC, Broekman MLD. Oversight in Surgical Innovation: A Response to Ethical Challenges. World J Surg. 2018 Mar 13.

S. Gupta · L. B. Fandino
Department of Neurosurgery, Computational Neurosciences Outcomes Center (CNOC), Brigham and Women's Hospital, Harvard Medical School, Boston, MA, USA

I. S. Muskens · A. F. C. Hulsbergen · M. L. D. Broekman (✉)
Department of Neurosurgery, Computational Neurosciences Outcomes Center (CNOC), Brigham and Women's Hospital, Harvard Medical School, Boston, MA, USA

Department of Neurosurgery, Haaglanden Medical Center, The Hague, The Netherlands

Department of Neurosurgery, Leiden University Medical Center, Leiden, The Netherlands
e-mail: m.broekman@haaglandenmc.nl; m.l.d.broekman@lumc.nl

the Belmont Report, a formative early guideline on clinical research ethics, innovative care is "practice that departs significantly from the standard or accepted," and such innovations warrant oversight, suggesting that operative innovations that depart from standard of practice should be regulated [1, 2]. Among other factors, the real-time nature of surgical innovation, uniqueness of each patient's anatomy, and emergence of cases make regulation more difficult in the operating room. Nonetheless, no standardized oversight mechanisms exist to protect neurosurgeons who seek to innovate and the patients they manage. This chapter discusses the ethical dilemmas associated with this lack of oversight and regulation and explores potential solutions for the current situation.

Oversight Mechanisms

Oversight mechanisms for surgical innovation have been posited. These range from surgical exceptionalism to centralized committees; the pros and cons of these are summarized in Table 4.1 [2, 3].

Surgical Exceptionalism

Surgical exceptionalism is characterized by self-regulation of an innovation by the surgeon without external overisight [3]. Its proponents generally argue that measuring and reproducing surgical technique, analyzing real-time decision-making, and making generalizations about anatomical variations make oversight impossible in the operating room.

Table 4.1 Summary of oversight mechanisms

Oversight level	Benefits	Drawbacks
Surgical exceptionalism	Surgeon knows patient's anatomy best, autonomy maintained, expedient	Susceptible to individual bias and COIs, interoperator inconsistency, no support for surgeons
Departmental	Surgeon knows patient's anatomy best, multiple opinions, autonomy maintained, expedient	Susceptible to institutional biases and COIs, interhospital inconsistency
Institutional	Multidisciplinary opinions, surgeon protected by legal and ethical expertise	Interhospital variability, professional independence may be compromised, moderately costly, and time-intensive
Regional/ national	Multidisciplinary opinions, sets precedent, no interoperator or interhospital variability	Subject to biases of the field, highly costly and time-intensive, assessment by evaluators removed from patient
IRB	Multidisciplinary opinions, standardized, transparent	Moderately costly and time-intensive, assessment by evaluators removed from patient

Departmental and Institutional Oversight

Departmental oversight may occur through informal conversation and conferences that integrate into existing educational conferences offered by a department. Such oversight is an efficient process which maintains the independence of the surgeon. It promotes collaboration and transparency within a department. However, it is subject to conflicts of interest and interdepartmental variations depending on the cultures of respective programs.

Institutional oversight builds on departmental conversations and conferences. Through institutional ethics committees (IECs) or surgical innovation committees (SICs), this oversight would include multidisciplinary perspectives toward an innovation. Institutions currently vary significantly in the composition and scope of their IECs. They have played larger roles in medical decision-making [4]. IECs may provide a more systematic approach toward an ethical dilemma. They may also offer alternative ethical, legal, and medical perspectives. However, the organization required in an IEC would slow the pace of innovation and may reduce the independence of the surgeon. They are also limited by variation between hospitals and may suffer from institutional biases.

Centralized Oversight

Oversight conducted by professional societies provides an increasingly centralized mechanism of regulation. While the American Medical Association's Council on Ethical and Judicial Affairs provides overarching ethical recommendations, no medical or surgical societies currently provide ethical committees to review ethical dilemmas with any regularity. Centralized oversight is currently hypothetical but would provide rigorous and methodological oversight while limiting individual and institutional biases. It could also offer multidisciplinary advice but would likely involve a costly and time-intensive review process.

Oversight in Formal Research Protocols

Operative innovation may also take place within the formal research paradigm regulated by institutional review boards (IRBs). Research should be undertaken with equipoise to generalize knowledge that will improve patient outcomes and aid surgeons in making decisions. Operative innovations in the non-research setting are similar and may even be precursors for research, but they differ because these innovations are tailored specifically to help individual patients without equipoise. IRB-approved innovative research may range from small prospective trials to the gold standard of clinical trials, randomized control trials (RCTs). However, inter-patient anatomic variation, ethical concerns with "sham" trial arms, relatively uncommon pathologies, and difference in operator ability limit the ability to conduct RCTs within neurosurgery. Therefore, most procedures are evaluated by single-operator/single-institution case series or individual case reports. The benefits of innovation as research include multidisciplinary assessment, minimization of COIs, and nationally standardized review processes. Clinical trials must be registered publicly as well, thus increasing their transparency [5]. IRB review may not be amenable to emergent cases and individual cases with significant anatomic variability due to its time-intensive and costly nature [6].

Ethical Justification for Formal Oversight

Oversight in innovation should strive to maintain the pillars of clinical ethics as extensively described in earlier work: scientific validity, risk-benefit ratio, informed consent, protection of vulnerable populations, justice, and conflicts of interest [7].

For the purposes of our framework, scientific validity and risk-benefit ratio are considered "scientific factors." Both of these factors involve statistical estimation with available objective scientific data and expertise. The scientific validity of an innovation depends largely on the quality of evidence that supports it, ranging from preclinical studies and untested expert opinion to clinical trials and meta-analyses. Defining the risk-benefit ratio of an innovation with surgical and patient-centered goals in mind is crucial in neurosurgery, where operations often require high risk and sacrificing one function for another. An assessment of risk must also include the surgeon's "learning curve" with a proposed innovation and an estimate of long-term risk [8].

We consider informed consent, protection of vulnerable populations, justice, and conflicts of interest (COI) "human factors" for this analysis because they involve more subjective assessment of communication, social justice, and bias. Informed consent for an innovative procedure should include explicit mention that it is a departure from standard of care, evidence to justify it, and the surgeon's experience applying it [9].

The practical justification for this division is that scientific factors are best judged by colleagues in the same field who are familiar and experienced with the relevant pathology and anatomy. Human factors, on the other hand, benefit from a more multidisciplinary approach that recognizes the legal and cultural contexts behind these ethical principles. These have been expounded in previous literature and are briefly summarized to motivate discussion for novel oversight mechanisms [9]. Vulnerable patients, including unconscious patients, patients in emergency conditions, children, prisoners, etc., require special considerations given their decreased abilities to advocate for themselves [9]. Care should be taken to avoid the exclusion or exploitation of vulnerable patients [10, 11]. Justice requires that the risks and benefits of an innovation are equitably shared. Financial COIs can occur when financial incentives promote the usage of certain surgical products. Nonfinancial COIs may include academic and public prestige.

Oversight as Quality Improvement

A quality improvement (QI) approach to oversight of surgical innovation could ensure both the authority of the surgeon and ethical practice. This approach aims to change the culture instead of the behavior of individuals and to accelerate innovation by providing an ethical and legal framework to neurosurgeons (Fig. 4.1). This

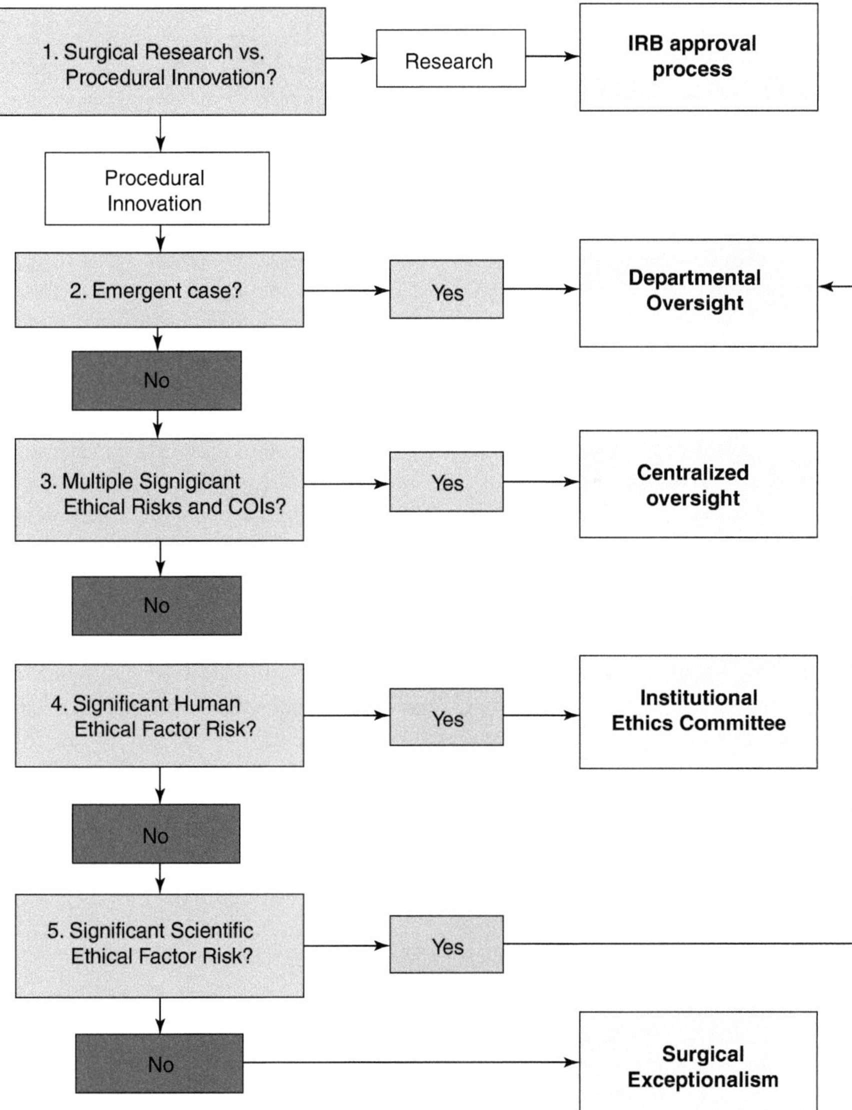

Fig. 4.1 Framework for the determination of appropriate level of oversight

framework is based on the Society of University Surgeons Surgical Innovations Project Team's position statement and recognizes different levels of innovation [12]. A form of registry was not included as outcomes from innovative procedures may be hard to compare due to specific anatomic variations between patients [12]. We suggest that this framework could aid the determination of the appropriate level of oversight when currently available tools identify the proposed procedure as an innovation [12, 13]. Neurosurgical innovations that pose greater risks to patients require a stricter form of oversight. However, the exact risk is hard to determine as many factors may be of influence, e.g., the experience of the surgeon, the emergence of the case, and available resources.

A QI approach requires outcomes that are measurable, like the surgical safety checklists and the complication-reducing Surgical Care Improvement Project [14]. This requires transparency, a willingness to share data, and preparedness to alter current practice. Possible methods to evaluate relevant outcomes could include surveys that determine the patients' understanding of the possible risks of the innovation, whether the case is adequately prepared for, and usability of the framework. More objective measures could be the number and severity of the neurosurgical innovations performed and their associated adverse events. Incorporated data analysis into the framework would improve the framework's content and delivery. Mutual trust is necessary to make QI initiatives by neurosurgeons a success and exportable to other phases and forms of surgical care [15].

It is important to distinguish surgical research from individual circumstances. This distinction is determined by its motivation; where the main goal of operative innovation is to optimize patient care, the main goal of research is to generate generalizable knowledge. The evaluation of a new technique or medical device is a form of research and thus requires oversight in the form of an IRB approval. A procedure that may be experimental in nature but is introduced by the neurosurgeon with the sole purpose to provide benefit to the patient is regarded as innovation for individualized clinical benefit. There are various forms of oversight in this setting and include surgical exceptionalism, informal discussion with colleagues, formal departmental conferences, IECs, and regional/national ethics committees (Table 4.1). The various ethical factors mentioned before determining the appropriate form of oversight. Examples are depicted in Table 4.2.

Table 4.2 Case examples of surgical innovations appropriate for different oversight levels

Oversight level	Illustrative examples for non-emergent cases
Surgical exceptionalism	New suture needle and string designed for intracranial microsurgery
Departmental	Use of a novel approved scope for endoscopic third ventriculostomy with limited experience
Institutional	Novel neuroendoscopic approach for an arteriovenous malformation near a comorbid meningioma in an obtunded patient
Regional/national	Endoscopic thoracic approach for thoracic hernia
IRB	Endoscopic third ventriculostomy for nonobstructive hydrocephalus

Surgical exceptionalism is appropriate when the innovative procedure poses no significant ethical challenges, and any form of regulation would put an unnecessary burden on the neurosurgical team. It may be hard to determine the risk-benefit ratio associated with the procedure beforehand, and the neurosurgeon must be able to adequately determine the appropriate level of oversight without being biased.

Departmental oversight may be appropriate when the innovative procedure poses scientific ethical challenges, but no human ethical challenges. In this scenario, there is no COI for the surgeon nor a risk of miscommunication between the surgeon and the patient. However, the evidence for the innovation may be limited and of low quality with poorly defined risk-benefit ratios as a result. The decision to innovate still lies with the neurosurgeon who knows the patient's anatomy and clinical history best, but the neurosurgeon's colleagues could help maximize the potential of the innovation by helping to define the associated efficacy and safety. It is unnecessary to involve IEC or centralized oversight committees as no human ethical factors are compromised.

IECs could be an appropriate form of oversight when human ethical factors are compromised, regardless of potential scientific ethical factors involved. An ICE could provide a well-funded answer to complex ethical challenges due to its multidisciplinary nature and internal discussion. The involvement of ethicists and lawyers could provide additional expertise to help neurosurgeon innovators presented with challenges with informed consent, protection of this vulnerable patient, and conflicts of interest. A disadvantage of IECs is the variation by institution regarding its role, scope, and makeup, which could result in inconsistent decision-making across various institutions. Standardization of IECs by neurosurgical and ethical societies is necessary to ensure that adequate expertise with neurosurgical ethical challenges is available and that decision-making is consistent. Collaboration between academic hospitals and nonacademic hospitals could enable access to IECs for all neurosurgeon innovators. It may not be possible to present an innovation to an IEC in the case of an emergency, and emergent innovation may require a lowered level of oversight. For example, a neurosurgeon presented with a severe neurotrauma in a child with a congenital anomaly that requires an unconventional decompression may not have the time to consult the IEC. The neurosurgeon will depend on discussion with colleagues and postoperative evaluation in this scenario.

A centralized oversight committee that is coordinated on a regional or national basis by neurosurgical societies may be the most appropriate form of oversight to introduce operative innovations with COIs and ethical challenges, both scientific and human. Clinical trials for institutional benefit and financial stakes in companies involved in the innovation could result in institutional COIs and may require further regulation [16]. The committees could mitigate these institutional COIs because of their centralized nature of these committees, but their formation would require current professional societies to be restructured to incorporate them. Funding for these committees could be acquired by patient advocacy organizations from state and national governments. Centralized oversight committees could maintain institutional independence and accelerate decision-making for time-dependent procedures by communication and integration with institutional committees through unbiased representatives. In the case of an emergent procedural innovation that would require oversight by a national committee, a lowered form of oversight may be required to ensure that the patient receives the necessary urgent care.

The decision to seek oversight currently lies with the neurosurgeon innovator, and the proposed framework is not meant to reduce the neurosurgeon's authority and ownership over their patients. Furthermore, the aim of the framework is to enable the neurosurgeon to innovate in an efficient and ethical fashion and to protect patients. Competence, integrity, humility, and consistency form the ethical pillars on which this QI framework is based. This framework would ideally be implemented in a culture of continuous self-improvement and learning to maximize its potential. Many challenges remain and range from the development of clinically applicable tools for identifying innovation and COIs, development of standardized ICEs, and integration of appropriate oversight with surgical care. We believe neurosurgical innovations can be performed consistently and ethically by implementing the appropriate form of oversight as suggested in our framework.

Conclusion

Current methods to address ethical challenges to operative innovation are inconsistent and open surgeons and patients to risk. Possible oversight mechanisms for operative innovation range from no oversight, associated with surgical exceptionalism, to formal IRB review—as for clinical trials. Certain oversight mechanisms may be well suited to regulate an attempt at innovation depending on the type and degree of pertinent ethical challenges to ensure the continued advancement of the field while protecting patients and supporting surgeons.

References

1. Bernstein M, Bampoe J. Surgical innovation or surgical evolution: an ethical and practical guide to handling novel neurosurgical procedures. J Neurosurg. 2004;100(1):2–7.
2. Research NCftPoHSoBaB. The Belmont report: ethical principles and guidelines for the protection of human subjects of research. Washington, DC: US Government Printing Office; 1978.
3. London AJ. Cutting surgical practices at the joints: individuating and assessing surgical procedures. 2006. https://www.cmu.edu/dietrich/philosophy/docs/london/London%2D%2DCutting SurgeriesatJoints.pdf.
4. Orlowski JP, Hein S, Christensen JA, Meinke R, Sincich T. Why doctors use or do not use ethics consultation. J Med Ethics. 2006;32(9):499–502.
5. Kim WO. Institutional review board (IRB) and ethical issues in clinical research. Korean J Anesthesiol. 2012;62(1):3–12.
6. Silberman G, Kahn KL. Burdens on research imposed by institutional review boards: the state of the evidence and its implications for regulatory reform. Milbank Q. 2011;89(4):599–627.
7. Emanuel EJ, Wendler D, Grady C. What makes clinical research ethical? JAMA. 2000;283(20):2701–11.
8. Healey P, Samanta J. When does the 'learning curve' of innovative interventions become questionable practice? Eur J Vasc Endovasc Surg. 2008;36(3):253–7.

9. Broekman ML, Carriere ME, Bredenoord AL. Surgical innovation: the ethical agenda: a systematic review. Medicine (Baltimore). 2016;95(25):e3790.
10. Hurst SA, Forde R, Reiter-Theil S, et al. Physicians' views on resource availability and equity in four European health care systems. BMC Health Serv Res. 2007;7:137.
11. Schwartz JA. Innovation in pediatric surgery: the surgical innovation continuum and the ETHICAL model. J Pediatr Surg. 2014;49(4):639–45.
12. Biffl WL, Spain DA, Reitsma AM, et al. Responsible development and application of surgical innovations: a position statement of the Society of University Surgeons. J Am Coll Surg. 2008;206(6):1204–9.
13. Blakely B, Selwood A, Rogers WA, Clay-Williams R. Macquarie Surgical Innovation Identification Tool (MSIIT): a study protocol for a usability and pilot test. BMJ Open. 2016;6(11):e013704.
14. Cevasco M, Ashley SW. Quality measurement and improvement in general surgery. Perm J. 2011;15(4):48–53.
15. SCOAP Collaborative, Writing Group for the SCOAP Collaborative, Kwon S, Florence M, et al. Creating a learning healthcare system in surgery: Washington State's Surgical Care and Outcomes Assessment Program (SCOAP) at 5 years. Surgery. 2012;151(2):146–52.
16. Emanuel EJ, Steiner D. Institutional conflict of interest. N Engl J Med. 1995;332(4):262–7.

The Ethics of the Learning Curve in Innovative Neurosurgery

5

Ludwike W. M. van Kalmthout, Ivo S. Muskens, Joseph P. Castlen, Nayan Lamba, Marike L. D. Broekman, and Annelien L. Bredenoord

Introduction

Due to innovation throughout the years, the modern and technologically advanced neurosurgical field that exists today is nearly unrecognizable from what it was decades ago. In particular, surgical outcomes of procedures have been tremendously improved by the introduction of new technologies, techniques, and procedures, such as bipolar cautery by Dr. Harvey Cushing, microsurgery by Dr. Gazi Yaşargil, and endoscopic surgery [1–3]. However, just because a new technology or technique is innovative does not mean it is an improvement over standard practice, and many innovations come with a host of ethical challenges. During the fledgling stages of the implementation of an innovative procedure, the associated learning curve presents one such ethical challenge. Many neurosurgeons may be relatively inexperienced with a brand-new procedure, and patients may face increased risks of complications as a result. How to balance the increased risk of complications faced

L. W. M. van Kalmthout · A. L. Bredenoord (✉)
Department of Medical Humanities, Julius Center for Health Sciences and Primary Care, University Medical Center Utrecht, Utrecht, The Netherlands
e-mail: a.l.bredenoord@umcutrecht.nl

I. S. Muskens · M. L. D. Broekman
Department of Neurosurgery, Computational Neurosciences Outcomes Center (CNOC), Brigham and Women's Hospital, Harvard Medical School, Boston, MA, USA

Department of Neurosurgery, Haaglanden Medical Center, The Hague, The Netherlands

Department of Neurosurgery, Leiden University Medical Center, Leiden, The Netherlands
e-mail: m.broekman@haaglandenmc.nl; m.l.d.broekman@lumc.nl

J. P. Castlen · N. Lamba
Department of Neurosurgery, Computational Neurosciences Outcomes Center (CNOC), Brigham and Women's Hospital, Harvard Medical School, Boston, MA, USA

© Springer Nature Switzerland AG 2019
M. L. D. Broekman (ed.), *Ethics of Innovation in Neurosurgery*,
https://doi.org/10.1007/978-3-030-05502-8_5

by patients with the overarching goal to innovate in order to improve outcomes for future patients is a major challenge that needs careful consideration. Here, we will examine the unique ethical challenges posed by the learning curve associated with an innovative procedure.

Definition of Learning Curve

In the literature, there seems to be no uniform definition of "learning curve," as it applies to innovative procedures. Some have described it as the gradual increase of knowledge and skill that comes with the repeated performance of the innovative procedure and perioperative patient care [4–6]. Others define "learning curve" as the gained knowledge and experience that is necessary for successful performance of the surgical procedure [7].

The influence of learning curves is recognized in several different settings of innovative surgery. Berstein et al. recognize three stages of innovation in which learning curves are relevant [8]. Starting with performing the procedure for the first time in stage one, followed by studies to prove safety and efficacy of an innovation that has shown to be beneficial to the individual patient in stage two, implementation of a procedure that has been proven to be beneficial and safe outside of the initial performing centers occurs in stage three.

Some authors discussed learning curves only in the setting of performing radically new procedures, such as in the first phase of the learning curve [9–11]. Interestingly, only two papers have described how evaluation of learning curves should be incorporated in a research setting, which would apply to the second phase of learning curves [7, 12]. Perhaps unsurprisingly, most authors describe the influence of learning curves only in the third phase, which consists of the implementation of the innovative procedure in general practice [13–16].

Authors' views of learning curves in the literature vary greatly, ranging from the opinion that they are an unavoidable part of surgical innovation [5, 17] to the view that they are a serious problem that needs to be addressed [5, 9, 13–16, 18–21]. Some have even described learning curves as a menace to patient safety [22].

Ethical Management of Learning Curves

Since innovative surgery by definition is initially performed by surgeons with little to no experience with the procedure in question, the associated learning curve could have unforeseen consequences. For instance, this inexperience could confound and complicate evaluation and interpretation of patient outcomes [12, 17, 19, 23]. Furthermore, adverse outcomes in the context of inexperience could result in reduced patient trust in the surgeon [5]. Since the scope of the risks of innovative procedures cannot always be fully defined, it is difficult, and in some cases impossible, for the surgeon to completely explain the

Table 5.1 Professional and technical requirements for each phase of innovation

Phase of innovation	Goal	Technical requirements	Professional requirements
Preclinical phase	Maximum preparedness for first procedure	• Train through simulation (e.g., cadaveric or computer models, etc.) • Evaluate relevant literature and operative videos • Shadow experts	
Clinical phase	Independent performance of the procedure	• Involve mentor for guidance • Review video postoperatively	Disclosure of relative inexperience during the informed consent procedure
Post-clinical phase	Maintain and enhance skills	• Participate in mentoring programs • Share experiences and learn from mistakes (e.g., on a conference) • Evaluate outcomes among peers	Evaluate personal experience and outcomes

risks involved in the procedure to the patient [13]. From an educational standpoint, since the attending surgeon in these cases has him or herself not completely mastered the procedure, surgical training of residents who are participating may not be completely effective either [4, 9]. How to deal with these and other aspects of learning curves in surgical innovation has been described by various authors, describing both technical and professional requirements (Table 5.1) [8, 4–29].

Technical Requirements

Surgeons have a moral obligation to train, and various training methods have been proposed to ensure technical competency of those performing an innovative procedure [5, 7–11, 14, 16, 19–21, 23, 26]. There are three different time periods when training is appropriate: (1) the preclinical phase, which involves preparation prior to the performance of the procedure on an actual patient; (2) the clinical phase, in which the procedure actually takes place; and (3) the post-clinical phase, in which proficiency of the surgeon is maintained [7].

Preclinical Phase

The purpose of the preclinical phase of training is to maximally reduce the risk to patient safety that is inherent from surgeon inexperience with a new procedure. In this phase, both cognitive and technical training are essential. In order to achieve adequate preparation, the use of in vivo, in vitro, computer, and cadaver models has been suggested in order to simulate human anatomy during training [5, 7, 9, 10, 16, 26]. Furthermore, the available literature and videos of similar cases (if available) should be studied extensively [6, 7, 11]. Finally, gaining first-hand experience from experts, when such experts are available, by visiting an specialized center or by doing a fellowship, may be valuable tools to understand the new procedure [6, 8, 9, 11, 16, 26]. It is crucial that surgeons are as prepared as possible for new procedures since the initial part of the learning curve is when they have the least experience and there is the greatest risk of adverse events for patients.

Clinical Phase

The clinical phase of training is comprised of the actual repeated performance of the new procedure, and it begins as soon as a surgeon performs the procedure on a real-life patient. Ideally, the first procedure is performed under the supervision of a surgeon with greater experience [9, 11, 26]. Although not always possible in the case of very new procedures, mentors could help to aid navigation, answer questions that may arise, and offer guidance via back-and-forth communication [16, 20]. Moreover, when experts in the technique are not available, any senior surgeon with more experience overall can offer technical guidance and advice in times of crises. Alternatively, some have suggested that reviewing operative videos may be sufficient in certain cases, such as when a new procedure is a slight variation on a familiar one [9]. Utilization of videos would also broaden the geographic reach of new procedures, allowing areas without a resident expert to receive training via technology. Finally, whenever possible, an expert should review the new surgeon's competency before he or she undertakes full independent performance of the procedure [8].

Overall, the ultimate goal is for the surgeon to be able to perform the procedure independently with the highest proficiency possible. To reach this point, multiple factors must facilitate the learning process for the surgeon, including strong mentorship, use of technology for training, as well as expert evaluation.

Post-Clinical Phase

After having gained enough experience with the new procedure so as to successfully perform the procedure independently, it is vital for the surgeon to maintain and enhance these acquired skills. Some have suggested that this should be carried out through a mentoring program [9]. A mentor could oversee and evaluate his mentee at various time points to optimize proficiency.

It is essential for the surgical community to share their experiences openly, exchange tips and advice, and problem-solve with one another based on the successes and failures they have faced when implementing a new procedure. This could be achieved through audits and conferences with the aim of discussing new procedures [20]. By learning from mistakes, identifying problems, and describing risks and limitations of the procedure based on experiences of a broad group of surgeons, outcomes could be improved more efficiently [20]. This continued improvement and expansion of accumulated knowledge comprises the post-clinical phase of training, with the hope being that this knowledge could be used to develop more accurate training modules to assist in the earlier phases of training [20].

Assessment of Learning Curve

These is no standardized method to assess the learning curve associated with innovative procedures. However, several ways to monitor the learning curve have been described [4, 8, 25]. These include the formation of specific oversight bodies. Some have suggested that these bodies could consist of one expert surgeon, whereas other have argued that multidisciplinary committees that could even oversee learning curves at multiple regional institutions would be superior [4, 6, 8, 25]. The goals of these committees could be to define standardized requirements for appropriate training, review fledgling innovative procedures, and provide accreditation [6, 8, 29].

Professional Requirements

Professional requirements of surgeons learning an innovative procedure include (1) obtaining adequate informed consent from patients and (2) honest communication of technical competency with peers [7–13, 16–24, 26–29]. During the acquisition of informed consent, transparent communication is essential in order to provide patients with accurate information about the relative inexperience of the surgeon performing the procedure [5, 11, 13, 16–19, 21, 22, 24, 27, 28]. This information could include a description of the success rate of the surgeon or other quantitative or qualitative forms of describing outcomes, both positive and negative [9, 24]. An honest disclosure becomes even more important when the surgeon performs the procedure for the first time [16]. Some view it as an obligation of the surgeon to evaluate their personal outcomes and reflect on their own skill and performance when deciding whether to get involved with an innovative procedure [4, 5, 9, 11, 24]. This can be achieved through accurate registration of outcomes with adequate follow-up, which then can be used to improve the training of other surgeons [5, 7, 9, 11, 12, 16, 18, 20, 24, 29].

Discussion

Even though there seems to be a lack of a clear definition of learning curves, they are inherent to surgical innovation and pose unique challenges. Ethical management of learning curves during the separate phases of innovation requires both technical and professional competencies.

Some have defined the learning curve only as a problem that arises when surgeons other than the original innovator start performing the procedure [13]. Whether or not this is the case, the experience of the primary investigator certainly could guide the learning process of other surgeons attempting to master the innovative procedure [7]. One could argue that the learning curve that attending surgeons face when performing an innovative procedure is similar to residents gaining experience with established procedures, which requires practice, mentorship, and supervision. As a result, the learning curves of residents could provide valuable insights that are also applicable to innovative surgery. There are several essential differences, however, between a resident's learning curve and the learning curve of a fully trained innovative surgeon. First of all, potential complications and outcomes of the procedures performed by the residents are well-defined, and the responsible attending has a wealth of experience with the procedures and can therefore step in and take charge should something go wrong. Beyond just the attending, during the training of residents, the whole team (including nurses and technicians) is experienced with the procedure, knows the typical outcomes, and has previously been involved in the training of residents. Conversely, during an innovative procedure, like the performing surgeon, the team is also inexperienced and unaware of the possible consequences, confounding and impeding the learning process which is more systematic for residents.

Therefore, in the case of performing a radically new procedure, the entire team has a moral obligation to prepare as much as possible to ethically manage the surgical learning curve. This is especially important since no earlier experience is available to guide decision-making intra- and perioperatively, which may be considered routine for established procedures. In this scenario, the IDEAL (idea, development, exploration, assessment, long-term follow-up) framework may prove helpful, as it describes clear steps that should be taken during development and implementation of innovative procedures [12]. This could be further aided by pre- and post-clinical training, which we deem as imperative to ensure patient safety by minimizing risks. Since each innovative procedure is unique, it requires a carefully tailored training program in order to achieve maximum preparedness. This could be attained through various forms of simulations and/or direct mentoring by an expert surgeon.

Currently, there is no standard way to assess learning curves of innovative procedures. Even though we think that specific oversight bodies as described above could play a role, another approach could be a critical review of a surgeon's performance (and that of the team) by the surgeon him or herself. This could take place before, during, and after preforming an innovative procedure and could be performed in collaboration with an expert or mentor. It is essential for a surgeon to have a proper understanding of his or her own limitations, and a framework that encourages self-review at every step of the process can optimize both the learning curve and ultimately patient outcomes, as well. Through self-reflection, a surgeon's errors and past complications are acknowledged, evaluated, and form a basis for future improvement. This may sometimes mean that a surgeon chooses to forego performing a certain procedure that he does not yet feel proficient in. This should be lauded rather than looked at as a sign of weakness so as to encourage open communication and continuous self-improvement in an environment where innovation is occurring so rapidly. One efficient way of achieving this may be through patient registries that promote adequate follow-up of all outcomes in order to evaluate not only safety but also progression of the surgeon along the learning curve. These registries may ultimately lead to shortening and optimization of the learning curve and result in prevention of adverse events.

Ethical management of learning curves requires also nontechnical skills as obtaining adequate informed consent. Transparent presentation of the experimental nature of the (un)known risks associated with the procedure is essential. A description of the surgeon's relative (in)experience is a necessity in order to meet the requirements of adequate informed consent: disclosure, decisional capacity, patient understanding of the information, voluntariness, and consent [21]. In reality, however, perhaps out of fear that the disclosure might confuse or distress the patient, present informed consent procedures probably do not meet these criteria [18]. Because fully disclosing the level of experience of a surgeon who is performing a new procedure may not cast that surgeon in the best light, some surgeons may avoid attaining informed consent in the proper manner and choose to leave out important information [18]. According to a survey among patients and surgeons, however, honest, descriptive disclosure of the risks and benefits and disclosure of whether the surgeon is performing the procedure for the first time appears to be the best approach [30].

Both adequate (self) assessment of learning curves and informed consent require specific ("soft") skills and an environment in which adverse events can be openly discussed with patients and peers. Self-reflection, verifiability, and honesty among surgeons form the foundations of such an environment.

Conclusion

Learning curves are associated with several challenges, including the increased risk of (unknown) complications, while surgeons master an innovative procedure. Ethical management of learning curves associated with innovative procedures in neurosurgery requires optimal preparation to balance these risks with potential benefits of a novel procedure. Surgeons have a moral obligation to train and need to meet several technical and professional requirements during the different phases of innovation.

References

1. Tew JM Jr. M. Gazi Yasargil: Neurosurgery's man of the century. Neurosurgery. 1999;45:1010–4.
2. O'Connor JL, Bloom DA. William T. Bovie and electrosurgery. Surgery. 1996;119:390–6.
3. Riskin DJ, Longaker MT, Gertner M, Krummel TM. Innovation in surgery: a historical perspective. Ann Surg. 2006;244:686–93.
4. McKneally MF. The ethics of innovation: Columbus and others try something new. J Thorac Cardiovasc Surg. 2011;141:863–6.
5. Healey P, Samanta J. When does the 'learning curve' of innovative interventions become questionable practice? Eur J Vasc Endovasc Surg. 2008;36:253–7.
6. Morreim H, Mack MJ, Sade RM. Surgical innovation: too risky to remain unregulated? Ann Thorac Surg. 2006;82:1957–65.
7. Neugebauer EA, Becker M, Buess GF, et al. EAES recommendations on methodology of innovation management in endoscopic surgery. Surg Endosc. 2010;24:1594–615.
8. Bernstein M, Bampoe J. Surgical innovation or surgical evolution: an ethical and practical guide to handling novel neurosurgical procedures. J Neurosurg. 2004;100:2–7.
9. Geiger JD, Hirschl RB. Innovation in surgical technology and techniques: challenges and ethical issues. Semin Pediatr Surg. 2015;24:115–21.
10. Moore FD. Ethical problems special to surgery: surgical teaching, surgical innovation, and the surgeon in managed care. Arch Surg. 2000;135:14–6.

11. Knight JL. Ethics: the dark side of surgical innovation. Innovations (Phila). 2012;7:307–13.
12. Hirst A, Agha RA, Rosin D, McCulloch P. How can we improve surgical research and innovation?: the IDEAL framework for action. Int J Surg. 2013;11:1038–42.
13. Angelos P. Ethics and surgical innovation: challenges to the professionalism of surgeons. Int J Surg. 2013;11(Suppl 1):S2–5.
14. Angelos P. The ethical challenges of surgical innovation for patient care. Lancet. 2010;376:1046–7.
15. Johnson J, Rogers W. Innovative surgery: the ethical challenges. J Med Ethics. 2012;38:9–12.
16. Miller ME, Siegler M, Angelos P. Ethical issues in surgical innovation. World J Surg. 2014;38:1638–43.
17. Jones JW, McCullough LB, Richman BW. Ethics of surgical innovation to treat rare diseases. J Vasc Surg. 2004;39:918–9.
18. Angelos P. Surgical ethics and the challenge of surgical innovation. Am J Surg. 2014;208:881–5.
19. Ethics Committee of American Society for Reproductive Medicine. Moving innovation to practice: a committee opinion. Fertil Steril. 2015;104:39–42.
20. McKneally MF, Daar AS. Introducing new technologies: protecting subjects of surgical innovation and research. World J Surg. 2003;27:930–4. discussion 4-5.
21. Wall A, Angelos P, Brown D, Kodner IJ, Keune JD. Ethics in surgery. Curr Probl Surg. 2013;50:99–134.
22. Morgenstern L. Position statement: surgical innovations. J Am Coll Surg. 2008;207:786. author reply.
23. VK T, Chow PK. An approach to the ethical evaluation of innovative surgical procedures. Ann Acad Med Singap. 2011;40:26–9.
24. Ethics ACo. ACOG Committee opinion no. 352: innovative practice: ethical guidelines. Obstet Gynecol. 2006;108:1589–95.
25. Dixon JB, Logue J, Komesaroff PA. Promises and ethical pitfalls of surgical innovation: the case of bariatric surgery. Obes Surg. 2013;23:1698–702.
26. Georgeson K. Surgical innovation and quality assurance: can we have both? Semin Pediatr Surg. 2015;24:112–4.
27. Lieberman JR, Wenger N. New technology and the orthopaedic surgeon: are you protecting your patients? Clin Orthop Relat Res. 2004:338–41.
28. Levin AV. IOLs, innovation, and ethics in pediatric ophthalmology: let's be honest. J AAPOS. 2002;6:133–5.
29. Strong VE, Forde KA, MacFadyen BV, et al. Ethical considerations regarding the implementation of new technologies and techniques in surgery. Surg Endosc. 2014;28:2272–6.
30. Lee Char SJ, Hills NK, Lo B, Kirkwood KS. Informed consent for innovative surgery: a survey of patients and surgeons. Surgery. 2013;153:473–80.

Innovation in Pediatric Neurosurgery: The Ethical Agenda

6

Bart Lutters, Eelco Hoving, and Marike L. D. Broekman

Introduction

The implementation of any surgical procedure ideally proceeds through various stages of development, from its first in-human application (innovation phase) to its validation by randomized trials (assessment phase) [1]. Particularly during the innovation phase, patients run the risk of being exposed to disproportionate risks, as treatment outcome cannot be predicted and ethical oversight is commonly lacking. Children are particularly vulnerable to potential hazards associated with an innovative surgical procedure, as they may lack the capacity to deliberate the risks and benefits associated with the innovation, are legally incapacitated, are subjected to prejudice, show deferential behavior toward adults, and often suffer from acute conditions for which no alternative therapy may be available [2]. In pediatric neurosurgery, patients commonly display most or all of the abovementioned vulnerabilities,

B. Lutters
Department of Neurosurgery, Erasmus University Medical Center,
Rotterdam, The Netherlands

Department of Pediatric Neurosurgery, Brain Center Rudolf Magnus,
University Medical Center Utrecht—Princess Máxima Center, Utrecht, The Netherlands

E. Hoving
Department of Pediatric Neurosurgery, Brain Center Rudolf Magnus,
University Medical Center Utrecht—Princess Máxima Center, Utrecht, The Netherlands

M. L. D. Broekman (✉)
Department of Neurosurgery, Computational Neurosciences Outcomes Center (CNOC),
Brigham and Women's Hospital, Harvard Medical School, Boston, MA, USA

Department of Neurosurgery, Haaglanden Medical Center, The Hague, The Netherlands

Department of Neurosurgery, Leiden University Medical Center, Leiden, The Netherlands
e-mail: m.broekman@haaglandenmc.nl; m.l.d.broekman@lumc.nl

© Springer Nature Switzerland AG 2019
M. L. D. Broekman (ed.), *Ethics of Innovation in Neurosurgery*,
https://doi.org/10.1007/978-3-030-05502-8_6

stressing the need to further optimize the informed consent process, to fasten the learning curve, and to provide for adequate ethical oversight.

Despite the need for ethical guidelines, concerns have been raised that stringent regulations and oversight would unnecessarily "hamper future progress of the field." Consequently, honest attempts to provide protection to children undergoing innovative neurosurgical procedures should be practical and not create additional hurdles for continuous innovation [1]. In this chapter, we aim to address the various ethical challenges associated with neurosurgical innovation in children and to propose the learning health system (LHS)—a system in which "knowledge generation is so embedded into the practice of medicine that it is the natural product of the healthcare delivery process and leads to the continuous improvement of care"—and its associated ethics framework by Faden and colleagues as a potential way to address these challenges without stifling neurosurgical innovation [3, 4].

Informed Consent

Over the past four decades, the concepts of clinical care and clinical research have been strictly separated; whereas clinical care is intended to benefit "a particular person in need of medical expert attention," the goal of clinical research is to obtain generalizable knowledge [5, 6]. This strict distinction has historically been prompted by various dubious research projects conducted by individuals as Robert Heath and Jose Delgado, in which research subjects were exposed to excessive hazards, without a reasonable chance to benefit from participation [7]. Since then, the premise that research should not be conducted on unwilling subjects, or on subjects somehow "tricked" into consenting, has been at the core of human research ethics.

Adequate informed consent consists of three components: disclosure, capability, and voluntariness. Hence, the informed consent process may be compromised when (1) inadequate information is provided by the surgeon regarding the risks and benefits associated with an innovative procedure, (2) the candidate-subject is incapable of weighing the risks and benefits associated with the procedure, or (3) consent is somehow forced upon the candidate-subject and, thereby, does not reflect his or her own opinion.

Throughout the informed consent process, the neurosurgeon is responsible for disclosing adequate information regarding the risks and benefits associated the innovative procedure based on the best available evidence. This poses a particular challenge to neurosurgical innovation in children, as the relatively "imprecise knowledge of functional anatomy and the high degree of anatomical variation and plasticity" make it hard to predict the risks and benefits associated with a novel neurosurgical intervention [7]. In pediatric epilepsy surgery, for example, patients commonly display a large degree of structural and functional rearrangement, making it hard to predict whether a certain innovative procedure may be safely performed without causing permanent deficits. Nevertheless, the neurosurgeon should always attempt to anticipate the consequences of performing the procedure, employing his or her expertise in the relevant disease and technical skill in similar procedures. Furthermore, adequate information should be provided about the innovative nature of the procedure and the lack of experience by the surgeon. Information should be provided at a level understandable to both the child and his or her proxy, which could be tested by separately asking the child to reproduce the information.

The informed consent process may be further impeded when the candidate-subject is incapable to consent or to deliberate the risks and benefits associated with the innovative procedure. At the most elementary level, a child may simply be unable to (verbally) express his or her wishes, which may be the case for infants, toddlers, and some mentally disabled children. In pediatric neurosurgery, these patients are particularly well represented, as congenital surgical procedures (e.g., spina bifida, hydrocephalus, and craniosynostosis surgery) make up a relatively large portion of pediatric neurosurgical practice. In addition, even when a child is able to communicate his or her wishes, he or she may not be capable of making a deliberate decision with regard to participating in a novel neurosurgical procedure. Indeed, "pediatric research subjects fall along a spectrum of decisional capacity," with some children clearly lacking the ability to weigh the risks and benefits associated with the innovation, whereas others may be mature enough to do so [2]. Hence, decisional capability needs to be established for each individual patient; a process in which the potential effects of neurological illness should also be taken into account, as these conditions may significantly alter patient judgment. When a child is ultimately considered incapable of providing adequate informed consent, parents or guardians will generally act as legal proxies. This legal subordination may, however, increase the child's vulnerability, as parents/proxies might not always act in the child's best interest and may potentially be influenced by personal interests. Such conflicting interests may also affect surgical decision-making, as the neurosurgeon may (despite his or her best intentions) potentially be misguided by career opportunities or prestige [8].

Lastly, even when adequate information is provided by the surgeon and the child has deliberately consented to the innovative procedure, the decision to consent may not necessarily reflect the child's own wishes. That is, a child may commonly tend to attribute much weight to the expectations of adults (e.g., parents and surgeons), even when this means setting aside personal convictions with regard to participating in the innovative procedure. Indeed, "the challenge is to devise a process that eliminates as much as possible the social pressures that a candidate-subject may feel," for instance, by leaving subject recruitment to independent healthcare workers with an interest in child psychology and by separately interviewing the child and his or her proxies [2]. Moreover, the cultural prestige surrounding the field of neurosurgery and loyalty toward their doctor may cause patients and proxies to simply accept an innovative procedure that they may, in fact, not agree upon [7].

Learning Curve

Patient outcome tends to improve as a surgeon gains experience in performing a certain procedure. Hence, the patients undergoing a novel neurosurgical procedure during the innovation phase may be exposed to higher risks and less benefits than patients operated on during later stages of development. Even though someone must be the first to undergo the procedure, the surgeon should always attempt to gain dexterity on animal models, human cadavers, and simulators prior to implementation [9]. Besides physical preparation, technical details of the procedure should be discussed among colleagues, thereby minimizing the occurrence of foreseeable complications. Moreover, the question whether a child—rather than an adult—should be the first to undergo a novel neurosurgical procedure must be addressed, as many neurosurgical conditions (e.g., brain tumors, epilepsy, and brain injury) affect both populations and children are generally more prone to perioperative complications, such as hypovolemic shock due to extensive blood loss.

The increasing use of deep brain stimulation (DBS) in children may serve as an illustrative example. Over the past decades, the surge of modern DBS has provided an effective treatment for adults suffering from various neurological and psychiatric disorders (e.g., dystonia, Parkinson, epilepsy, Tourette syndrome, and obsessive-compulsive disorder). Only after its safety and efficacy had been well established in the adult population, DBS was carefully adopted in children, ultimately targeting the same neuroanatomical regions that were used in adults. Consequently, when DBS was first applied in the pediatric population (mostly for cases of severe dystonia), neurosurgeons had already gained considerable experience in the procedure, and appropriate neuroanatomical targets had already been selected. The introduction of an established neurosurgical procedure in pediatric subjects should, however, not be taken lightly; the plasticity of the developing brain may have a significant influence on the efficacy of DBS, and the long-term effects of chronic DBS remain to be elucidated [10]. So while the adult use of DBS has allowed neurosurgeons to practice the procedure prior to its application in pediatric subjects, children may still be exposed to significant risks as results from adults cannot simply be extrapolated.

Lack of Oversight

Ethical and regulatory guidelines are essential to protect children from disproportionate risks potentially associated with innovative neurosurgical procedures. Except for the strict regulations regarding novel neurosurgical implants, however, neurosurgical innovation generally takes place with relatively little oversight [7, 9]. This has previously been justified by *surgical exceptionalism*, "the view that the somewhat exceptional ethical or regulatory status of surgery is justified by the unique nature of surgery" [11]. As a consequence, moral responsibility for implementing a novel neurosurgical procedure is primarily placed upon the neurosurgeon, who—despite his or her sincere efforts—may not be sensible to the child's decisional (in)capability or deferential behavior and may not always be free of conflicting interests or social prejudice.

Neurosurgical innovation occurs along a spectrum, ranging from minor deviations from standard practice (e.g., the use of longer screws for cranioplasty in a patient with mild cranial hyperostosis) to the introduction of radically innovative procedures (e.g., the implantation of 3D printed skull in a patient suffering from generalized cortical hyperostosis) [8, 9]. These different types of innovations clearly require different mechanisms of oversight; whereas minor modifications are generally considered inherent to surgical practice and would, therefore, require little ethical oversight, most authors agree that some form of formal oversight is needed with regard to the implementation radical innovations. The use of external review committees has previously been suggested for this category of innovation [9].

An additional problem may arise from the lack of oversight regarding innovations that occur in an emergency setting. As innovation in pediatric neurosurgery frequently takes place in an acute setting (e.g., traumatic brain injury)—in which there is no time for adequate informed consent—children may sometimes be exposed to risks to which neither they nor their parents or guardians have consented to. Even though consent may be sought following the procedure, progressive neurological deterioration or neurosurgical manipulation may have significantly altered the patient's judgment [7]. Consequently, additional protection should be provided to children subjected to radically innovative procedures in an emergency setting. Ethical review committees specifically dedicated to emergency innovations may serve to establish the risks and benefits associated with the innovation outside of the emergency setting [12].

Discussion

Innovation in pediatric neurosurgery gives rise to various ethical challenges. Even though these challenges emphasize the need for ethical guidelines, such guidelines may potentially interfere with honest attempts to provide current and future patients with the best available neurosurgical care. The learning health system (LHS) aims to bridge the traditional divide between research and care by stimulating alternatives to large randomized trials, fostering universal data sharing and improving public and professional understanding of the importance of contributing to evidence-based medicine [4]. To help ensure that research activities within an LHS are conducted in an ethically acceptable fashion, Faden and colleagues have proposed a moral framework that significantly departs from contemporary clinical and research ethics by addressing the moral obligation to address unjust inequalities and the moral obligation to contribute to continuous learning activities [3].

The moral obligation to address unjust inequalities by Faden and colleagues supports the traditional notion that burdens associated with a learning activity should not fall disproportionately on patients already socially or economically disadvantaged [3]. The obligation also extends beyond the traditional conception by imposing an obligation "to direct learning activities toward aggressive efforts to reduce or eliminate unfair or unacceptable inequalities in the evidence base available for clinical decision-making" [3]. That is, healthcare professionals ought to make sure that learning activities are not solely directed toward the common majority but should also target patient minorities. Currently, children undergoing neurosurgery are infrequently subjected to clinical investigation (due to concerns regarding the vulnerability), which has led to a relative paucity of evidence in this patient population.

The moral obligation to address unjust inequalities—in combination with the alternative study designs proposed by the learning health system—may stimulate ethically acceptable learning activities directed toward pediatric neurosurgical patients [3, 4].

The moral obligation to contribute to continuous learning activities applies to both healthcare workers and patients. Whereas the responsibility of healthcare workers to provide optimal patient care—based on the best available evidence—has been commonly addressed, the moral obligation to actively contribute to this evidence base deviates from traditional clinical and research ethics [3]. This is based on the assumption that healthcare workers have such a "unique access to and control over clinical care and health information" that it would be immoral not to harness this position to advance the quality and justice of the healthcare system [3]. The obligation to contribute to learning activities may stimulate healthcare workers to report and share healthcare data among institutions. This may accelerate the generation of evidence by means of cluster randomized trials and observational treatment comparisons, thereby potentially overcoming the ethical challenges associated with the randomized controlled study design in pediatric neurosurgery. In addition to the moral obligation placed on healthcare workers, patients are thought to have a similar obligation to contribute to continuous learning activities [3]. This obligation is based on the principle that those who receive the benefits from an effective LHS—a system which requires "continuous access to information from as much patients as possible to be efficient"—have the moral responsibility to contribute to its advancement [3]. Consequently, when no harm is to be reasonably expected from a certain learning activity (e.g., observational treatment comparisons), patients may have the moral obligation to participate in the activity. This would, obviously, not mean that children have a moral obligation to participate in potentially hazardous neurosurgical innovations without the need for informed consent but rather provides an ethical justification for the universal use and sharing of anonymous patient data to continuously assess pediatric neurosurgical procedures.

In conclusion, we believe that the LHS and its associated ethics framework may help to overcome the challenges associated with innovation in pediatric neurosurgery. Even though innovation will remain indispensable for progress of the field, ethically justified efforts to come up with alternatives to large randomized trials may take away some of the professional reluctance to set up neurosurgical trials. This may ultimately accelerate the transition from the relatively unregulated innovation phase to subsequent phases of surgical development, thereby allowing for systematic outcome assessment and ethical oversight earlier on in the implementation process.

Conclusion

Neurosurgical innovation in children may offer long-lasting benefits but also gives rise to various ethical challenges. First, the informed consent process may be comprised when a child lacks decisional capacity or shows deferential behavior toward adults. A second challenge arises from the limited experience with the innovative procedure, which may expose the child to excessive risks as compared to subsequent patients (stressing the need to gain dexterity on animal models, human cadavers, simulators, and adults). Thirdly, the lack of regulation and oversight in neurosurgical innovation puts children at risks of undue exploitation by those in authority. We believe that the learning healthcare system and associated ethics framework by Faden and coworkers may help to overcome these ethical challenges by fostering the search for alternative study designs while providing adequate protection to pediatric neurosurgical patients.

References

1. McCulloch P, Altman DG, Campbell WB, Flum DR, Glasziou P, Marshall JC, et al. No surgical innovation without evaluation: the IDEAL recommendations. Lancet. 2009;374:1105–12.
2. Kipnis K. Seven vulnerabilities in the pediatric research subject. Theor Med Bioeth. 2003;24:107–20.
3. Faden RR, Kass NE, Goodman SN, Pronovost P, Tunis S, Beauchamp TL. An ethics framework for a learning health care system: a departure from traditional research ethics and clinical ethics. Hast Cent Rep. 2013;43:16–27.
4. Olsen L, Aisner D, McGinnis JM. The learning healthcare system: workshop summary (IOM Roundtable on Evidence-Based Medicine). Washington DC: National Academies Press; 2007. p. xi–xvi.
5. Kass NE, Faden RR, Goodman SN, Pronovost P, Tunis S, Beauchamp TL. The research-treatment distinction: a problematic approach for determining which activities should have ethical oversight. Hast Cent Rep. 2013;43:4–15.
6. Miller FG. Revisiting the Belmont Report: the ethical significance of the distinction between clinical research and medical care. APA Newslett Philos Med. 2006;5:10–4.
7. Ford PJ. Vulnerable brains: research ethics and neurosurgical patients. J Law Med Ethics. 2009;37:73–82.
8. Schwartz JA. Innovation in pediatric surgery: the surgical innovation continuum and the ETHICAL model. J Pediatr Surg. 2014;49:639–45.
9. Broekman ML, Carrière ME, Bredenoord AL. Surgical innovation: the ethical agenda: a systematic review. Medicine. 2016;95:1–5.
10. Lipsman N, Ellis M, Lozano AM. Current and future indications for deep brain stimulation in pediatric populations. Neurosurg Focus. 2010;29:1–7.
11. London AJ. Cutting surgical practices at the joints: individuating and assessing surgical procedures. Ethical guidelines for innovative surgery. Hagerstown: University Publishing Group; 2006.
12. Nwomeh BC, Waller AL, Caniano DA, Kelleher KJ. Informed consent for emergency surgery in infants and children. J Pediatr Surg. 2005;40:1320–5.

Conflicts of Interest in Neurosurgical Innovation

Aislyn C. DiRisio, Ivo S. Muskens, David J. Cote,
William B. Gormley, Timothy R. Smith, Wouter A. Moojen,
and Marike L. D. Broekman

Introduction

Conflicts of interest (COI) are inherent components of innovative ventures and can be particularly powerful in neurosurgery. COI are often unavoidable and occur when any individual or group has conflicting loyalties. While some COI go without impact, others have the potential to cause harm to patients, undermine trust between a patient and his or her surgeon, or adversely affect the quality of innovation. It is important for innovators to critically evaluate their loyalties, collaborate with others

This chapter is based in: DiRisio AC, Muskens IS, Cote DJ, Babu M, Gormley WB, Smith TR, Moojen WA, Broekman ML. Oversight and Ethical Regulation of Conflicts of Interest in Neurosurgery in the United States. Neurosurgery. 2018 May 30.

A. C. DiRisio · D. J. Cote · W. B. Gormley · T. R. Smith
Department of Neurosurgery, Computational Neurosciences Outcomes Center (CNOC),
Brigham and Women's Hospital, Harvard Medical School, Boston, MA, USA
e-mail: david_cote@hms.harvard.edu

I. S. Muskens · M. L. D. Broekman (✉)
Department of Neurosurgery, Computational Neurosciences Outcomes Center (CNOC),
Brigham and Women's Hospital, Harvard Medical School, Boston, MA, USA

Department of Neurosurgery, Haaglanden Medical Center, The Hague, The Netherlands

Department of Neurosurgery, Leiden University Medical Center, Leiden, The Netherlands
e-mail: m.broekman@haaglandenmc.nl; m.l.d.broekman@lumc.nl

W. A. Moojen
Department of Neurosurgery, Haaglanden Medical Center, The Hague, The Netherlands

Department of Neurosurgery, Leiden University Medical Center, Leiden, The Netherlands

Department of Neurosurgery, Haga Teaching Hospital, The Hague, The Netherlands

© Springer Nature Switzerland AG 2019
M. L. D. Broekman (ed.), *Ethics of Innovation in Neurosurgery*,
https://doi.org/10.1007/978-3-030-05502-8_7

to ensure adherence to ethical practice, and mitigate the effects of their COI on patient care and innovation. By managing COI responsibly, neurosurgeons and others involved in neurosurgical innovation can maintain respect of patient autonomy, ensure beneficence of treatment, avoid maleficence to maintain public trust, and improve the quality of innovation.

The medical device industry is critical for funding, enabling, and promoting innovation in neurosurgery. Neurosurgeons' involvement in the medical device industry provides critical insight for new medical devices and imaging technologies, and thus COI is rightfully ubiquitous in neurosurgery. This can become problematic if business interests of a company impact a neurosurgeon's decisions regarding clinical care and innovative practice. The role of a neurosurgeon has naturally leant itself to autonomy and self-governance, as well as the responsibility to act in the best interest of the patient. Nevertheless, the strong dependence on technology and robust culture of innovation outside the formalized structure of randomized controlled trials (RCTs) in neurosurgery can create powerful financial and professional incentives for innovation and use of technologies. In addition to financial incentives, the desire to advance a career in academia, improve financial outcomes, publish papers, and gain status all create biases and, at times, unnoticed COI that affect innovative pursuits and the use of technology in clinical practice. While these are important to the success and advancement of the field, it is critical that care is taken to responsibly manage the natural COIs that develop as a result of these innovative pursuits so as to always protect patient safety. Here, we discuss and evaluate COI that affects innovation, its use, and its dissemination in neurosurgical practice.

Clinical Practice

There are indirect sources of COI that can influence the decisions made by all surgeons. In the world of academia, professional incentives exist to publish on novel techniques in order to improve academic standing and status. Similarly, these can also give the physician an opportunity to improve their financial compensation, expand their referral volume, and increase operative productivity. It has even been argued that having industry representatives present in the operating room can cause favorability toward a particular company based on bias toward the representative [1]. This practice, however, is not only common but is critical to improve the use of implants and ensure patient safety. Thus, this creates the risk that even the indirect relationships with industry can compromise patient care [2, 3]. Consideration could be given, however, to how routine clinical practice is carried out and how to promote ethical innovation in neurosurgery.

It is often difficult to distinguish between clinical care and innovation of thought in surgery. "Surgical exceptionalism" is the idea that surgical innovation is not conducive to oversight and that the surgeon is best able to make decisions on behalf of the patient [4]. While research and medical devices has obvious forms of oversight, innovation of thought and innovative procedures rarely have oversight. As was the case in the development of endoscopic endonasal meningioma surgery, these procedures typically involve a gradual change in practice over time [5].

Additionally, neurosurgeons are increasingly faced with the need to make complex decisions; in addition to survival, other factors such as quality of life, invasiveness of a procedure, and recovery time are becoming increasingly important [6]. For novel procedures and devices, less is known about these features and complication rates, because there is often less data available [7]. While personal experience is a critical source of evidence to inform surgical decision-making, especially in the absence of data on a novel procedure, this nonetheless allows more room COI to influence decision-making. The surgeon is well equipped to discuss the risks and benefits of a procedure, yet informed consent can be easily affected by a physician's biases, experience, and financial COI, all of which affect physician estimates of risk, what procedures to perform, and what devices to use [8]. This unquestioned autonomy, in addition to the complexity of surgical decision-making, can make surgeons especially susceptible to COI, which is often subtle or indirect. Patients, of course, expect that physicians will guide them to make decisions in a way that is ethically sound and without the influence of COI [9].

At neurosurgical meetings, device manufacturers frequently sponsor discussions about products and surgical dilemmas. These talks may also influence the clinical judgment of attendees, especially if financial incentives are present or if COI is not adequately disclosed to allow the reader to assess the content in context. Outside the realm of neurosurgery, similar sponsorship from drug companies was shown to have a favorable effect on drug prescribing patterns regardless of whether the physicians remembered who sponsored the study [10]. These patterns highlight how pervasive COI is. It is important to not only report COI but also to include information on the role taken by the funder in presented research, to allow the reader to judge the quality and independence of studies and form their own conclusions about the presented results, if desired.

Ties to Industry

Of all fields, neurosurgery may be one of the most robust in terms of innovation. The input and feedback provided by physicians to device manufacturers is crucial in the development of medical devices, and in fact 20% of medical device patents have at least one physician-inventor [11]. Payments are made to physician-innovators for their expertise and time spent developing and testing new medical technology. Direct ties to industry may bring financial benefits to neurosurgeons in exchange for innovative thinking and advancing clinical practice and thus help to incentivize progress and compensate for personal risk. In 2014, the Open Payments Database, which was created by the Affordable Care Act to make payments to physicians publically available, recorded over $1,000,000,000 in payments to neurosurgeons from medical device and pharmaceutical companies. Surgical involvement in the device industry is also valuable to clinical care by allowing neurosurgeons to become well versed in the utilization of novel devices and learn about the technology directly from the company [12, 13]. All of these ties are critical to ensure that the field moves forward.

Though the ultimate purpose of the medical device industry are to produce products to help the patients, the underlying goals to create a profit and run a business venture are naturally misaligned with the ideal aims of academic research: to provide unbiased, evidence-based answers. If not handled appropriately, these goals may lead to poorly designed trials, study enrollment that is inadequate or not generalizable, biased data interpretation, or insufficient reporting of adverse events [14]. This was shown in the case of the recombinant human bone morphogenetic protein (rhBMP) spinal implant, in which the results of industry-sponsored publications were misleading compared to other estimates. In this case, COI was thought to be one factor leading to inadequate reporting of adverse events. In addition to being an example of unethical practice, events such as these provide a disservice to medicine and the public by allowing for publication of potentially misleading results [15]. Financial COI is an inevitable component of neurosurgical progress, and thus it is critical that these COI are managed in a way that is ethically sound and clinically practical, in order to maintain quality.

In addition to their role as an innovator, neurosurgeons can be involved in the early implementation of novel medical devices, consult with industry, sit on advisory boards for device companies, and receive industry funding for their own research. These roles are critical to improve the quality of innovation from the medical device industry yet also create the possibility for bias and favorability toward a particular company [1, 16]. Additionally, vendors are frequently present during operations to provide on-the-spot input in the use of novel hardware and surgical instruments and input from surgeons providing device manufacturers with valuable clinical insight to improve technology and ensure that the products are patient-oriented [1, 6, 17, 18]. Royalty payments greater than $5000 awarded to physicians who are among the first to use devices from Medtronic Spinal and Biologics, Stryker Spine, and Synthes Spine were expected to pass 13 million dollars in 2010 [19]. While an inherent component of the industry, the fear with these payments is that it may provide undue pressure on a physician to opt for a particular device [16].

In the United States, legal disclosure of financial COI was not required of physicians until 2010, when the Sunshine Act was enacted as part of the Patient Protection and Affordable Care Act (PPACA). With this act, physicians became required to report their consulting fees, compensation, or company ownership in companies with at least one product covered by Medicare. The goal of this was to prevent inappropriate power of industry over clinical judgment [20]. While the websites for the Sunshine Act are publicly searchable, the data available are difficult to interpret, and there is a lack of public knowledge about the sites and what the COIs mean for patient care. If the patient is unaware of the reporting altogether, legal disclosure does little to reduce the influence of COI in practice [1]. Additionally, patients who are aware of the database admittedly do not fully understand the extent of relationships between the device industry and physicians [16]. While one might think that COI could cloud clinical judgment, other COI may be unrelated to a case at hand or may even be an indication of expertise on a procedure or device. Similarly, patients find gifts that physicians commonly accept from industry to be immoral, yet are not necessarily in favor of stronger government regulations [21]. Thus, while public disclosure is an important step in legal reporting of COI, it does not have a major effect on day-to-day patient care and may need to be supplemented with policies to address when additional consent and patient education is specifically needed.

Concerns over COI highlight the need for a comprehensive informed consent protocol. These could preserve patient autonomy by ensuring that they have at least a minimal level of knowledge regarding their neurosurgeon's ties with industry and whether the device they are having implanted is innovative in nature or an off-label use. Ensuring that the patient is kept informed about the procedure will, at the very least, hold the surgeon accountable and encourage a conversation over what this means, if the patient desires. Regardless of the source of the oversight, it is nonetheless important that patients are protected from the influence of COI on clinical practice and research.

Even if a patient is made aware of the innovative nature of a procedure and physician COI, they can sometimes fall victim to the assumption that novel is necessarily better [6]. Therapeutic misconception is the idea that patients do not understand the difference between treatment and research, and thus believe that their providers will always act in their best interests, as has been shown in trials in which 100% of patients expect positive results [22, 23]. This is further complicated in surgical fields, where the boundary between innovation of thought and clinical practice is less clear. Furthermore, for innovative procedures with limited available information about the long-term risks, as can be the case in neurosurgery given the small numbers of patients, a truly informed discussion about risks and benefits may not always be possible.

Neurosurgical Journals

While industry plays an important role in funding neurosurgical research, the influence that industry has on the neurosurgical literature goes beyond enabling research to happen. It has been shown that medical research funded by industry is more likely to report positive outcomes than research without industry funding, for example [24]. In the case of the rhBMP spinal implant, COI may have been one of many factors leading to insufficient reporting of adverse events. COI was reported insufficiently and inconsistently, rendering the study difficult to understand in the context of financial ties with industry [15].

In 1984, *The New England Journal of Medicine* became the first journal to formally require disclosure of author conflicts of interest because of both the inevitability of industry-academia relationships and the importance of maintaining public trust [25]. Following their lead, requiring disclosure of COI among all authors has become more common, with 70% and 90% of biomedical journals requiring authors to report of nonfinancial and financial COI, respectively [26]. Inconsistencies exist in reporting, however, and include the lack of a definition as to what constitutes COI and selective disclosure for authors without COI makes the reporting of COI (or lack thereof) difficult to interpret [15, 27]. Disclosure allows the reader to understand the research in the context of other factors, and thus consistent, transparent reporting of COI could allow the reader to evaluate the reputability of the methods and results for themselves. Therefore, clear definitions on what constitutes a COI could be developed by neurosurgical journals to maintain the integrity of the literature. Additionally, author COI could be included in articles, even if the authors have no disclosures, to allow the reader to interpret the results in context.

Unlike the reporting of COI for authors, it is far less common for COI of journal editors to be disclosed [28]. Among biomedical journals, less than 40% of journals require editors to report COI [26]. One study found that at least 29% of editors of five leading spine journals had a financial conflict of interest reported at meetings, yet this was determined to be a highly conservative estimate due to an inability to trace COI for about half of the editors. Of these editors with a financial COI, 76% of their financial relationships were with major medical device companies, and 42% had more than $10,000 disclosed in a source other than the journal [28]. Given the assumption of objectivity in the peer-review process, a process in which reviewers and editors have been described as the "gatekeepers" of science [29], disclosure of COI among editors can help to maintain the legitimacy of peer-reviewed publications and maintain public trust. By making editor COI publically available on journal websites for readers to easily find and assess for themselves, readership or journals will be better equipped to judge the research quality.

Some journals have addressed this through transparent policies regarding when editors and reviewers must recuse themselves from the review of a particular manuscript, including when the manuscript in question includes authors from their home institution or was funded by a company they have financial ties to [30]. Policies such as these are important because they provide a way to protect the quality of the literature without compromising the peer-review process. The *Journal of Neurosurgery* and their related journals also require that editors submit annual disclosure statements and allow for their editors to excuse themselves if unable to be impartial in the peer-review process.

More effort can be made to regulate the quality of studies submitted. From ethical incidents have come some examples of stricter oversight in the editorial process [31], yet regulation has not become the standard in medical journals [28], and similar proactive policies could be made more widespread to protect patients. Furthermore, innovative surgical devices do not have to be compared against the "gold standard," both to be published and to be approved by the FDA. This has caused harm to patients, such as in the case of the interspinous process device (IPD) [32]. For these devices, supposed safety and efficacy was based on single-arm retrospective studies for 30 years, until prospective studies and randomized controlled trials eventually evaluated them and found them and found faults. While this may not always be possible, requiring or encouraging studies submitted to compare the novel innovation to the standard of care could ensure that the literature is relevant. Additionally, requiring demonstration of methods to reduce the effects of bias to publish in surgical journals could help improve trust with readership, especially for studies receiving industry funding. Ideally, these could be designed by investigators without a financial stake in the results. Because of the small numbers of patients seen in neurosurgical practice and the nature of surgery, it is especially important that all adverse events and long-term outcomes are reported, so data can be pooled and reevaluated to further assess quality of innovation. Additionally, another study found that 24% of devices in some areas were recalled for safety reasons 5–10 years after their introduction, highlighting a need for monitoring. Maintaining the quality of the published literature and allowing the reader to understand the study in the context of COI will allow for improved safety in the application of the literature to clinical practice and will improve the integrity of the literature.

Despite efforts to improve the quality of the medical literature, it is important to note that the "grey literature" and social media have a profound effect on both patients and physicians and is far less regulated. Disclosure is not standard in these forms of communication, and there is no effort to eliminate the effects of bias. It is increasingly important to critically evaluate all information we receive and provide patients with the resources that they need to make informed decisions, in order to improve the quality of care.

Institutions

Hospitals can also have COI that influence their ability to provide the best care for their patients. By adapting new technology, a hospital can both improve revenue and status [7]. New technology is often expensive, however, and thus the investment gives the hospital incentive to advertise widely and increase utilization. The data available on new technologies is often incomplete, biased, or conflicting. In neurosurgery, the use of intraoperative MRI significantly increases the expense of treatment for the patient because the high cost of implementation and prolonged operative time, yet many neurosurgeons feel that the improved imaging brings substantial benefit. The data on whether this improves outcomes remains a subject of debate [33]. Although many patients benefit from this advancement, many other patients may have no need for a technology that provides them with marginal benefit at an increased cost but may view the innovation as superior or mistake the marginal benefit as worthwhile. Adapting new technology is nonetheless important to move the field of neurosurgery forward but further complicates surgical decision-making as neurosurgeons are tasked with helping patients make the best decisions.

The incentive to provide the best possible treatment and the belief that new, expensive, innovative approaches necessarily improve outcomes affecting the patient may both influence their choice when it comes to decisions on where to receive treatment and which treatment to receive. It could be argued that this is especially true in patients with particularly devastating diseases as is seen in neurosurgery, who are willing to travel to academic centers and receive multiple consultations to find the best options for them. Thus, hospitals may have an additional incentive to implement innovative, expensive technologies before there is conclusive supporting evidence.

Additionally, institutional oversight of neurosurgical practice is one way to improve ethical, patient-centered practice, improve patient autonomy in research, and help patients understand the disclosures that they receive. Though disclosure policies exist at the majority of medical schools, only 1% of institutions surveyed required disclosure to research subjects, and many policies used vague language and inadequately defined terminology, thus leaving much of reporting up to the physician [34]. If surgeons are to remain autonomous, patients expect accountability and sound decisions, regardless of COIs [35]. Awareness of the effects of bias and disclosure does little to change behavior [36], further supporting the need for stricter institutional enforcement of COI policy. Furthermore, patients have admitted to not necessarily being able to interpret disclosures [21], and thus it is critical to give patients the tools to understand COI and assess the associated risks and benefits [37, 38] rather than bypassing patient involvement in their own care. A more robust means of reporting, as enforced by institutions, may help maintain autonomy of neurosurgical patients.

Institutional policies need clear definitions within their policies and requirements for complete transparency with all financial relationships to ensure adequate disclosure to patients. Although the Sunshine Act made financial COI public, patients could be better informed of how to use this and what it means through the consent process. One such proposal on the institutional level is to prohibit inventors from being involved in clinical testing for companies for which they invented devices for or have a consulting relationship with [12]. This has been criticized as being too strict as to stifle innovation [39] and has since been relaxed, yet it also prevents unintentional bias and increases the likelihood of obtaining results that are both reproducible and generalizable. Other suggestions to reach the same results have included giving some investigators read-only access to research data and involving others without a financial COI in the study design and data interpretation [40]. Because of the unique nature of surgery, however, it is also recommended that multiple surgeons, especially those without ties to the innovation, are involved in implanting a device or performing a technique for the first time. This could ensure generalizability of results and improve adherence to evidence-based practice.

Conclusion

COI are ubiquitous and inevitable in innovative practice. While both financial and nonfinancial COI can affect neurosurgery, these collaborative efforts are also valuable to promote growth and progress. Even awareness of COI and reporting does not necessarily change practice, and thus it is important to always think about practices alongside other collaborators without bias and look at practices from an ethical perspective to maintain patient-centered, evidence-based practice among neurosurgeons.

References

1. Johnson J, Rogers W. Joint issues—conflicts of interest, the ASR hip and suggestions for managing surgical conflicts of interest. BMC Med Ethics. 2014;15:63.
2. Gelberman RH, Samson D, Mirza SK, Callaghan JJ, Pellegrini VD Jr. Orthopaedic surgeons and the medical device industry: the threat to scientific integrity and the public trust. J Bone Joint Surg Am. 2010;92(3):765–77.
3. Shufflebarger HL. Surgeons, societies, and companies: ethics and legalities. Spine. 2001;26(18):1947–9.
4. London AJ. Ethical guidelines for innovative surgery. Hagerstown: University Publishing Group; 2006.
5. Bernstein M, Bampoe J. Surgical innovation or surgical evolution: an ethical and practical guide to handling novel neurosurgical procedures. J Neurosurg. 2004;100(1):2–7.
6. Angelos P. Ethics and surgical innovation: challenges to the professionalism of surgeons. Int J Surg. 2013;11(Suppl 1):S2–5.
7. Miller ME, Siegler M, Angelos P. Ethical issues in surgical innovation. World J Surg. 2014;38(7):1638–43.
8. Brunelli A, Pompili C, Salati M. Patient selection for operation: the complex balance between information and intuition. J Thorac Dis. 2013;5(1):8–11.
9. Namm JP, Siegler M, Brander C, Kim TY, Lowe C, Angelos P. History and evolution of surgical ethics: John Gregory to the twenty-first century. World J Surg. 2014;38(7):1568–73.
10. Spingarn RW, Berlin JA, Strom BL. When pharmaceutical manufacturers' employees present grand rounds, what do residents remember? Acad Med. 1996;71(1):86–8.
11. Chatterji AK, Fabrizio KR, Mitchell W, Schulman KA. Physician-industry cooperation in the medical device industry. Health Aff (Project Hope). 2008;27(6):1532–43.
12. Stossel TP. Regulating academic-industrial research relationships—solving problems or stifling progress? N Engl J Med. 2005;353(10):1060–5.
13. Kimmelstiel C. Restrictions on interactions between doctors and industry could ultimately hurt patients. J Vasc Surg. 2011;54(3 Suppl):12s–4s.
14. Nelsen LL, Bierer BE. Biomedical innovation in academic institutions: mitigating conflict of interest. Sci Transl Med. 2011;3(100):100cm126.
15. Carragee EJ, Hurwitz EL, Weiner BK. A critical review of recombinant human bone morphogenetic protein-2 trials in spinal surgery: emerging safety concerns and lessons learned. Spine J. 2011;11(6):471–91.
16. Fadlallah R, Nas H, Naamani D, et al. Knowledge, beliefs and attitudes of patients and the general public towards the interactions of physicians with the pharmaceutical and the device industry: a systematic review. PLoS One. 2016;11(8):e0160540.
17. Rajaratnam A. Current trends in the relationship between orthopaedic surgeons and industry. J Bone Joint Surg Br. 2009;91(10):1265–6.
18. Brown J. Growth and innovation in medical devices: a conversation with Stryker chairman John Brown. Interview by Lawton R. Burns. Health Aff (Project Hope). 2007;26(3):w436–44.
19. Aarabi BC, Chesler D, Maulucci C, Blacklock T, Alexander M. Dynamics of subdural hygroma following decompressive craniectomy: a comparative study. Neurosurg Focus. 2009;26(6):E8.
20. Smith SW, Sfekas A. How much do physician-entrepreneurs contribute to new medical devices? Med Care. 2013;51(5):461–7.
21. Camp MW, Gross AE, McKneally MF. Patient views on financial relationships between surgeons and surgical device manufacturers. Can J Surg. 2015;58(5):323–9.
22. Appelbaum PS, Roth LH, Lidz C. The therapeutic misconception: informed consent in psychiatric research. Int J Law Psychiatry. 1982;5(3–4):319–29.
23. Constantinou M, Jhanji V, Chiang PP, Lamoureux EL, Rees G, Vajpayee RB. Determinants of informed consent in a cataract surgery clinical trial: why patients participate. Can J Ophthalmol. 2012;47(2):118–23.

24. Bailey CS, Fehlings MG, Rampersaud YR, Hall H, Wai EK, Fisher CG. Industry and evidence-based medicine: believable or conflicted? A systematic review of the surgical literature. Can J Surg. 2011;54(5):321–6.
25. Relman AS. Dealing with conflicts of interest. N Engl J Med. 1984;310(18):1182–3.
26. Bosch X, Pericas JM, Hernandez C, Doti P. Financial, nonfinancial and editors' conflicts of interest in high-impact biomedical journals. Eur J Clin Investig. 2013;43(7):660–7.
27. Probst P, Huttner FJ, Klaiber U, Diener MK, Buchler MW, Knebel P. Thirty years of disclosure of conflict of interest in surgery journals. Surgery. 2015;157(4):627–33.
28. Janssen SJ, Bredenoord AL, Dhert W, de Kleuver M, Oner FC, Verlaan JJ. Potential conflicts of interest of editorial board members from five leading spine journals. PLoS One. 2015;10(6):e0127362.
29. Park IU, Peacey MW, Munafo MR. Modelling the effects of subjective and objective decision making in scientific peer review. Nature. 2014;506(7486):93–6.
30. Gottlieb JD, Bressler NM. How should journals handle the conflict of interest of their editors?: who watches the "watchers"? JAMA. 2017;317(17):1757–8.
31. Carragee EJ, Hurwitz EL, Weiner BK, Bono CM, Rothman DJ. Future directions for The spine journal: managing and reporting conflict of interest issues. Spine J. 2011;11(8):695–7.
32. Moojen WA, Bredenoord AL, Viergever RF, Peul WC. Scientific evaluation of spinal implants: an ethical necessity. Spine. 2014;39(26):2115–8.
33. Witiw CD, Nathan V, Bernstein M. Economics, innovation, and quality improvement in neurosurgery. Neurosurg Clin N Am. 2015;26(2):197–205, viii.
34. McCrary SV, Anderson CB, Jakovljevic J, et al. A national survey of policies on disclosure of conflicts of interest in biomedical research. N Engl J Med. 2000;343(22):1621–6.
35. Christian F, Pitt DF, Bond J, Davison P, Gomes A, Bond J. Professionalism—connecting the past and the present and a blueprint for the Canadian Association of General Surgeons. Can J Surg. 2008;51(2):88–91.
36. Cain DM, Detsky AS. Everyone's a little bit biased (even physicians). JAMA. 2008;299(24):2893–5.
37. Thompson DF. Understanding financial conflicts of interest. N Engl J Med. 1993;329(8):573–6.
38. White AP, Vaccaro AR, Zdeblick T. Counterpoint: physician-industry relationships can be ethically established, and conflicts of interest can be ethically managed. Spine. 2007;32(11 Suppl):S53–7.
39. Bailey M. Harvard Medical School puts strict ethics rules under microscope. The Boston Globe. 2015.
40. Institute of Medicine Committee on Conflict of Interest in Medical Research, Education, and Practice. The National Academies Collection: Reports funded by National Institutes of Health. In: Lo B, Field MJ, editors. Conflict of Interest in Medical Research, Education, and Practice. Washington (DC): National Academies Press (US) National Academy of Sciences; 2009.

Joseph P. Castlen, David J. Cote, and Marike L. D. Broekman

The Economics of Neurosurgical Innovation

As in much of medical innovation, new treatments and procedures in neurosurgery are often extremely expensive to develop. As such, the costs of innovation are sometimes the rate-limiting step in the process of innovation [1]. The development of new therapies and procedures, however, is intimately tied to the idea of progress in neurosurgery and is a driving force in the improvement of patient outcomes. This conflict between rising costs of innovation and the inherent need to innovate gives rise to some of the ethical questions involving the funding of neurosurgical innovation.

In 2015, over $150 billion was spent on medical research and development (R&D) [2]. The private sector medical research industry contributes almost twice as much to R&D funding as all other sources combined, with over $100 billion spent [2]. The next highest contributor was the US federal government with almost $36 billion, over 80% of which came from the National Institutes of Health (NIH) [2]. Smaller but still significant funders of research were academic institutions with $8.5

J. P. Castlen · D. J. Cote
Department of Neurosurgery, Computational Neurosciences Outcomes Center (CNOC), Brigham and Women's Hospital, Harvard Medical School, Boston, MA, USA
e-mail: david_cote@hms.harvard.edu

M. L. D. Broekman (✉)
Department of Neurosurgery, Computational Neurosciences Outcomes Center (CNOC), Brigham and Women's Hospital, Harvard Medical School, Boston, MA, USA

Department of Neurosurgery, Haaglanden Medical Center, The Hague, The Netherlands

Department of Neurosurgery, Leiden University Medical Center, Leiden, The Netherlands
e-mail: m.broekman@haaglandenmc.nl; m.l.d.broekman@lumc.nl

© Springer Nature Switzerland AG 2019
M. L. D. Broekman (ed.), *Ethics of Innovation in Neurosurgery*,
https://doi.org/10.1007/978-3-030-05502-8_8

billion, philanthropic and patient groups with $5 billion, and state and local government with another $1.5 billion [2].

Neurosurgical innovation requiring the development of medical therapies (e.g., vaccine trials for glioblastoma) or major new technologies (e.g., spine implants) can cost anywhere from tens of thousands to tens of millions of dollars [1, 3–5]. Although such innovation is clearly quite costly to fund, it is important to note that this is not always the case. Sometimes, "innovation" in neurosurgery amounts to changes in familiar surgical techniques or slight modifications to common instruments used during surgery. Even when innovation does not require much funding in the way of R&D, it can still cost a substantial amount in the way of hospital bills, professional fees, postoperative follow-up, and other costs associated with transporting, rooming, and boarding patients who live far from treatment facilities.

Before a procedure is deemed reasonable and necessary for treatment of a particular condition, insurance companies refuse to reimburse for the costs of providing care. Typically, when insurance companies do not pay for procedures, the responsibility for payment falls on the patient. Novel procedures, however, present a special case in which a patient is sometimes being asked to pay for a service that is not known to be successful. Here, we will address this and other ethical issues that surround each potential funding source for innovation in neurosurgery.

Examining Sources of Funding

Innovators, who often already have busy clinical and administrative schedules, must spend time writing grants, corresponding with industry representatives and reviewing research proposals in order to obtain funding for their innovation. Furthermore, they may face limitations on how they can spend awarded money. Funding for innovative procedures and treatments can come from a variety of sources, public and private, each of which comes with potential ethical problems.

Patient Self-Payment

Although it is standard practice for patients to "self-pay" for medical procedures that insurance companies will not cover, there are causes for concern in allowing patients to self-pay for unproven treatments. The first is that since potential complications cannot always be reasonably predicted, there is a possibility that costs will be much higher than anticipated. This is true for most surgical procedures, but the unknown factor is even greater for novel procedures. Even if there are no complications, asking a patient to pay for a treatment when there is no scientific evidence that it will be helpful is tantamount to asking a patient to gamble on his or her own well-being, except that in this case there are not always well-defined odds.

Another issue is the potential that patients could be selected based on their ability to pay, leading to a bias in which only wealthy patients are offered novel treatments. Ideally, patients would be selected for novel procedures solely based on their clinical status, since selection of only certain types of patients could confound interpretation of outcomes. Still, an argument could be made that if patient self-payment is only needed on a temporary basis until insurance companies begin reimbursing costs, then the effects of initially treating only wealthy patients become less severe. Furthermore, if a treatment draws patients from far distances, then the ability to afford travel expenses could select out some patients anyway, adding an extra layer of selection even when costs of treatment are covered.

In any case, it is fundamentally better to select patients without regard to ability to pay based on the commonly accepted principle of justice in medical ethics [6]. The American Medical Association's own Code of Medical Ethics states, "A physician shall support access to medical care for all people" [7]. This principle is certainly intended foremost to safeguard against discrimination, whether on racial, religious, or other grounds; however, it can be applied in this case as well if a patient's ability to pay is interpreted as an indicator for his or her socioeconomic status.

Still, it is not difficult to sympathize with the hypothetical patient who can afford to fund his or her innovative procedure but is not allowed to do so on ethical grounds, instead being forced to rely on potentially lengthy, tedious, and bureaucratic processes to obtain funding from elsewhere. The justice argument can seem tenuous in light of the fact that some innovative procedures may only need to be performed a handful of times before obtaining insurance approval and also given that some patients' access to typical, non-innovative healthcare is already limited by their ability to pay.

For these reasons, we do not recommend a categorical ban on patient self-payment for innovative procedures. Instead, self-payment by patients should only be considered after reasonably exhausting all other potential funding sources, and hospital ethics committees should be consulted before allowing it to happen. There may be cases in which it is acceptable for patients to fund innovation, but they are likely few and far between, and this practice should be avoided.

Insurance Involvement

Insurance companies do not explicitly fund innovation; however, they determine which innovative treatments are widely available by covering the costs of the treatments. Insurance approval of new procedures is not easily obtained despite the fact that it may actually help to decrease costs and increase patient quality of life in the long term [8]. Usually, FDA approval or a clear consensus within the medical community is necessary for insurance companies to begin covering a new treatment [9]. When insurance companies do not cover new procedures or treatments, large costs fall onto patients, reintroducing some of the previously discussed problems with patient self-payment for innovative treatments. Additionally, this impedes further study of the innovative procedures since presumably fewer patients will have access to the said treatment.

Industry Funding

As previously mentioned, the private sector devotes by far the most money in the United States to medical R&D. Simply put, medical innovation, including in neurosurgery, would not be possible today without the financial support of pharmaceutical companies and medical device manufacturers. In the literature, some have described partnerships with industry as beneficial to neurosurgical research and have advocated for strengthening relationships between academic centers and industry researchers [10]. The large amount of money that goes into innovation from industry undoubtedly helps advance medical research and development of new devices and techniques; however, it also creates a reliance that subjects innovators to certain ethically problematic pressures.

Collaboration between clinicians and industry researchers is unavoidable – clinicians often require financial support from industry which in turn requires clinicians since they are the gateway for industry researchers to access patients. It is not difficult to identify potential problems with these collaborations. For instance, some neurosurgeons may have financial interests in the success of technologies if they are invested in companies supporting said technologies, creating a conflict of interest that could cause bias in the conducting or interpretation of research.

This also joins together two entities which have fundamentally different missions. Nonprofit universities and teaching hospitals typically have stated missions involving education and the advancement of patient care, and while industry mission statements also often place an emphasis on patient well-being or putting patient care first, the nature of a for-profit company necessitates that research is ultimately viewed a means to an end (turning a profit) rather than an end in itself (improving patient care). This raises the concern that research funded by industry may be biased and that new surgical innovations may not actually be necessary or, in a dystopian scenario, are actually detrimental to patients.

Perhaps epitomizing this concern are multiple studies which have found that industry-funded research is more likely to report positive results than nonindustry-funded work [11–13]. Although there are multiple explanations for this that do not involve bias in research interpretation or design – for instance, maybe industry representatives are better at selectively funding studies that are more likely to succeed – the obvious inference is that the research is somehow either being influenced or presented in a biased manner not consistent with the traditional view of academic medical innovation. This fact implies that it would be better if fewer studies were funded this way, though not necessarily at the expense of slowing innovation.

Philanthropy

Perhaps at the opposite end of the spectrum from financially motivated industry-driven funding is philanthropic funding. Whether it comes in the form of individual donations, patient support groups, charitable foundations, or other sources, there is a significant amount of money (about $5 billion in the United States) that is contributed to medical R&D from philanthropic sources [2]. The literature is limited in its analysis of individual donors funding research, but ethical questions persist [14].

In addition to freeing innovators from many potential conflicts of interest, philanthropic contributions to innovation require little time input on the part of the innovator in terms of writing grants, completing paperwork, etc. Donations may be made toward the research fund of a particular surgeon or researcher, toward a general cause, or toward a specific treatment under investigation. These types of donations can be made either directly to an innovator, to the institution where the innovator is based, or to a charitable foundation which then distributes funds according to its mission.

Philanthropy is a unique form of funding because those who give money do not normally place any limitations on how it can be spent, other than the reasonable assumption that it is put to good use. They do not usually personally stand to profit, other than perhaps fulfilling a personal desire to help cure a disease or make the world a better place, and physicians are likewise not involved in any conflicts of interest as a result of these donations. It is not likely to ever become the primary method of funding innovation, but from an ethical standpoint, it is not very prone to issues for neurosurgical innovators.

Academic/Institutional Support

Given their educational missions, it is no surprise that universities are a substantial contributor to medical innovation in the United States. Like federal funding for research, the motivation behind this funding is theoretically in the public interest rather than for any financial gain. In practice, institutional funding is also often motivated by a desire on the part of the institution to profit off of patents, boost their reputation, and attract desirable researchers and physicians. Despite this and the fact that they do not contribute as much financially as industry or the federal government, universities and hospitals are uniquely positioned to play a significant role in advancing innovation.

Hospitals, as an employer of a large number of clinicians/innovators, are often the physical location of innovation. Innovators, as such, are positioned to turn to hospitals for funding in the event that a particular innovation cannot be feasibly funded through other means discussed here. Even when it is not directly funding innovation, hospitals, as the nexus of innovation, can help facilitate its implementation. For instance, they could waive facility fees for patients being seen in hospital clinics or hospital costs for inpatients. While this certainly is not expected of hospitals, they have the power to make these decisions in special cases to support innovation carried out by their employees.

In order to encourage innovation (potentially resulting in financial gain), to advance their educational missions, or to increase the prestige of the institution, some universities and hospitals directly fund innovation [15, 16]. Since the pool of applicants to grants from individual hospital or universities is relatively small, innovators with ideas that would likely not win larger national grants may be more inclined to apply for these funds. Furthermore, institutions may fund promising projects within their own network in order to retain budding innovators whose careers are not yet far enough along to win larger outside grants. In this way, universities and hospitals, with their limited resources, may act as stimulators of innovation rather than primary financiers of it.

Government Grants

In the United States, the second-largest source of funding for medical research and development is the government [2]. Of this money, hundreds of millions went to neurosurgical research and billions to broader neurologic research [17]. This is despite a relative decline in funding for surgical research compared to nonsurgical research [18, 19]. One potential reason for this is a decline in the number of grant proposals being submitted by neurosurgeons due to their busy schedules and the bureaucracy involved in preparing these proposals [1, 4].

Although it is far from the largest supporter of medical research, the government still provides a substantial amount of money to researchers. One advantage of publicly funding research is that the impetus for funding is ostensibly purely for the public good. In other words, there is not much concern that financial concerns will influence the outcomes of research. It is true that political pressures could affect the distribution of money to particular research topics; however, the degree to which this occurs is likely less than the forces at work in the private sector [20, 21]. There are also professional pressures on innovators who win grants to "publish or perish," to produce short-term outcomes, and to win more and more grants as their careers progress.

Of course, arguments can be made as to whether it is the role of the government to fund scientific and medical research. Some argue that it is a burden on society and that it should be curtailed or ended; however, the positive effect of government money for innovators is undeniable [22]. Although the grant application and review process requires significant time investment on the part of the principle investigator, the process does help to ensure that only well thought out proposals are funded. Indeed, in neurosurgery, NIH funding is tied to a higher scholarly impact in published research [23, 24].

The Big Picture

In addition to the sources discussed here, there are other alternative sources of funding. For instance, the ancillary costs of innovation (e.g., patient travel costs for treatment) may be covered by crowdfunding websites. These sites certainly help patients, but in addition to patient privacy and other concerns, they have the potential for abuse given they are unregulated and that recipients of donations are seldom held accountable for the money they receive [25, 26]. Like patient self-payment, crowdfunding medical expenses incurred during innovation should be avoided if possible [27].

There is also a distinction between what type of innovation is being funded through each source. Most industry funding goes to translational research, while government grants (which only constitute a fifth of all medical R&D money) fund almost half of basic science research [28]. Historically, the government has been the primary funder of basic science research, while industry focused more on clinical trials and translational research [28]. Although basic science research does not lead directly to innovation, it does lay the groundwork for innovative research to take place.

There is not a single funding source that drives innovation in neurosurgery. Due to the ethical concerns discussed here, patients should be shielded from as much financial responsibility as possible when undergoing innovative treatments. In particular, the cost of the treatment itself should not be charged to patients, and ideally travel costs would be covered as well. Meanwhile, innovators should be proactive in avoiding conflicts of interest. In cases where this is not possible, conflicts should be clearly declared so as not to give the illusion of undue influence.

No matter where the money comes from, innovation will inherently come with some degree of pressure for certain results, if not just from the hope that an innovation will increase the standard of care for a particular disease. While innovators should avoid conflicts, funding sources should also be cognizant of and resist the temptation to create conflicts, whether they are financial, professional, or some other types of pressure to get certain results from innovators. Virtually all funding sources have a stated mission of improving patient care, and they must not lose sight of this mission. Accurate and honestly reported research results will ultimately produce the best outcomes for patients.

References

1. Schnurman Z, Kondziolka D. Evaluating innovation. Part 2: development in neurosurgery. J Neurosurg. 2016;124(1):212–23.
2. America Research. U.S. Investments in Medical and Health Research and Development, 2013-2016. 2016. p. 3–6.
3. Henaine AM, Paubel N, Ducray F, et al. Current trends in the management of glioblastoma in a French University Hospital and associated direct costs. J Clin Pharm Ther. 2016;41(1):47–53.
4. Leuthardt EC. What's holding us back? Understanding barriers to innovation in academic neurosurgery. Surg Neurol. 2006;66(4):347–9; discussion 349.
5. Witiw CD, Nathan V, Bernstein M. Economics, innovation, and quality improvement in neurosurgery. Neurosurg Clin N Am. 2015;26(2):197–205, viii.
6. Beauchamp TL, Childress JF. Justice. In: Principles of biomedical ethics. 7th ed. New York: Oxford University Press; 2013. 459 p.
7. AMA Code of Medical Ethics. American Medical Association; 2001.
8. Lakdawalla D, Malani A, Reif J. The insurance value of medical innovation. National Bureau of Economic Research Working Paper. 2015.
9. Tsoi B, Masucci L, Campbell K, Drummond M, O'Reilly D, Goeree R. Harmonization of reimbursement and regulatory approval processes: a systematic review of international experiences. Expert Rev Pharmacoecon Outcomes Res. 2013;13(4):497–511.
10. Firlik AD, Lowry DW, Levy AJ, Hirsch RC. The neurosurgeon as innovator and entrepreneur. Neurosurgery. 2000;47(1):169–75; discussion 175–67.

11. Bhandari M, Busse JW, Jackowski D, et al. Association between industry funding and statistically significant pro-industry findings in medical and surgical randomized trials. CMAJ. 2004;170(4):477–80.
12. Ridker PM, Torres J. Reported outcomes in major cardiovascular clinical trials funded by for-profit and not-for-profit organizations: 2000-2005. JAMA. 2006;295(19):2270–4.
13. Lundh A, Lexchin J, Mintzes B, Schroll JB, Bero L. Industry sponsorship and research outcome. Cochrane Database Syst Rev. 2017;2:MR000033.
14. Pratt B, Hyder AA. Fair resource allocation to health research: priority topics for bioethics scholarship. Bioethics. 2017;31(6):454–66.
15. Chase D. Hospital-based strategic venture funds to spark innovation. In Forbes. Online; 2012.
16. Orrell K, Yankanah R, Heon E, Wright JG. A small grant funding program to promote innovation at an academic research hospital. Can J Surg. 2015;58(5):294–5.
17. National Institutes of Health. Estimates of Funding for Various Research, Condition, and Disease Categories (RCDC). February 10, 2016 2016.
18. Mann M, Tendulkar A, Birger N, Howard C, Ratcliffe MB. National institutes of health funding for surgical research. Ann Surg. 2008;247(2):217–21.
19. Rangel SJ, Efron B, Moss RL. Recent trends in National Institutes of Health funding of surgical research. Ann Surg. 2002;236(3):277–86; discussion 286–77.
20. Reardon S. Lobbying sways NIH grants. Nature. 2014;515(7525):19.
21. Kaiser J. Biomedical research policy. NIH funding shifts with disease lobbying, study suggests. Science. 2012;338(6104):181.
22. Kealey T. End Government Science Funding. The Kato Institute; 1997.
23. Svider PF, Husain Q, Folbe AJ, Couldwell WT, Liu JK, Eloy JA. Assessing National Institutes of Health funding and scholarly impact in neurological surgery. J Neurosurg. 2014;120(1):191–6.
24. Hauptman JS, Chow DS, Martin NA, Itagaki MW. Research productivity in neurosurgery: trends in globalization, scientific focus, and funding. J Neurosurg. 2011;115(6):1262–72.
25. Ozdemir V, Faris J, Srivastava S. Crowdfunding 2.0: the next-generation philanthropy: a new approach for philanthropists and citizens to co-fund disruptive innovation in global health. EMBO Rep. 2015;16(3):267–71.
26. Snyder J, Mathers A, Crooks VA. Fund my treatment!: a call for ethics-focused social science research into the use of crowdfunding for medical care. Soc Sci Med. 2016;169:27–30.
27. The Lancet Oncology. Mind the gap: charity and crowdfunding in health care. Lancet Oncol. 2017;18(3):269.
28. Mervis J. Data check: U.S. government share of basic research funding falls below 50%. In: ScienceMag. American Association for the Advancement of Science; 2017.

Part II

Payment for and Right to Innovation in Neurosurgery

Public Pressure for Neurosurgical Innovation

9

David J. Cote

Introduction

Public advocacy has long played a prominent role in shaping healthcare policy and biomedical research priorities. During the early years of the HIV/AIDS epidemic, large-scale protests against pharmaceutical companies led to early release of HIV medications and expanded access through compassionate use, as well as more public funding for HIV research and education [1]. Public advocacy has played a key role in the advancement of cancer research, with well-known fundraising and awareness campaigns like the Jimmy Fund and the Susan G. Komen Foundation critically influencing healthcare policy and research throughout their history [2–4].

Neurosurgical innovation is driven forward by a variety of factors, the most prominent of which is a desire to improve patient outcomes. Ancillary motivating factors include opportunities for professional and academic advancement, financial gain, and an abstract desire for the field to progress. These motivations mostly come from within the field of neurosurgery and thus can be considered methods of "self-governance" of innovation—ways that the field itself encourages its own development and progress [5, 6]. There are sources of pressure for neurosurgical innovation, however, that come from outside this community. These include, for example, public advocacy campaigns, public fundraising and philanthropy, and pressure from the public for specific innovation, each of which can play a role of its own in motivating neurosurgical innovation and progress.

While these campaigns often have noble goals, there are times where public pressure for innovation can be problematic, particularly in cases where the benefits of the innovative procedures being sought are unknown. In neurosurgery, for example, there has been a major push to expand access to minimally invasive

D. J. Cote (✉)
Department of Neurosurgery, Computational Neurosciences Outcomes Center (CNOC), Brigham and Women's Hospital, Harvard Medical School, Boston, MA, USA
e-mail: david_cote@hms.harvard.edu

© Springer Nature Switzerland AG 2019
M. L. D. Broekman (ed.), *Ethics of Innovation in Neurosurgery*,
https://doi.org/10.1007/978-3-030-05502-8_9

techniques for many surgeries, simply because they are sought by patients, as well as to advance stem cell procedures, despite a lack of data showing that these are effective (and some data showing they may be dangerous) [7–9].

Public pressure for neurosurgical innovation can push the goals of clinical research and practice in both positive and negative directions. In this chapter, we aim first to identify some of the ways that the public influences research and clinical trends in neurosurgery and then to evaluate the ethical issues surrounding them. As public advocacy continues to grow in the age of social media and the Internet, neurosurgeons will need to combine their own clinical training and the concerns of the public responsibly, to achieve outcomes satisfying for both parties.

Public Funding, Fundraising, and Philanthropy

The power of the public dollar often goes a long way in shaping the goals of biomedical research in general, and neurosurgical research and innovation are not exempt from these pressures. Basic research, translational research, and clinical research all require financial support. Often, clinical innovations also require financial support, with external funding supporting costs of research and development or clinical implementation of an innovative procedure. While much of the funding for biomedical research in the United States comes from federal agencies, other sources include private donors and industry [10].

Federal Funding

In the United States, funding for research and innovation frequently comes from federal agencies like the National Institutes of Health (NIH) [11, 12]. Federal agencies are well-equipped to review applications, manage funds, and award grants on merit and are supported by large administrative structures, many employees, and relatively stable funding from the federal government. In return for this stable funding, researchers are expected to dedicate significant portions of their career to research and innovation and to provide return on investment through publications, presentations, and education of trainees.

For neurosurgical innovation, funding from the federal government remains a common and well-regulated solution that is not the primary focus of this chapter. Surgical innovations often require funding for the support of device research and development, patient enrollment, and staff compensation, but these proposed procedures are often not well-suited to funding from organizations like the NIH. Often, innovation in neurosurgery begins with just a few patients or is the result of a small modification of an existing procedure that would not merit large-scale funding [9, 13]. In these cases, innovation is excluded from the regulatory structures and influence of organizations like the NIH [5]. Because less funding for these projects comes from well-established federal sources, alternative sources of funding, like philanthropists or patient advocacy organizations, have a heightened impact.

Private Donors

In some cases, private philanthropic donors can exert influence over research aims. In many cases, these donors have a personal connection to a disease, with a personal history of illness or a friend or family member affected by a particular disease. In these cases, philanthropic donors are often motivated by a desire to provide support for clinical advancement in the treatment of that particular disease and often work with a specific surgeon or small group of surgeons, such as a neurosurgical department.

While the goals of these donations are noble, the resulting relationships can occasionally be problematic. Receiving donations for clinical innovation can put neurosurgeons in the difficult position of providing innovative clinical services while simultaneously being subject to significant conflict of interest [6, 14–16]. Overbearing involvement by a philanthropic donor can result in overstatement of the potential of a treatment by the neurosurgeon, application in situations that may not be clinically appropriate, or selective provision of the particular clinical innovation to patients related to the donor who provides funding.

Funding for this type of treatment has even expanded to social media. After being diagnosed with an astrocytoma, one family from the United Kingdom raised over £175,000 to pay for proton therapy for their daughter that would not be provided in short enough time under their existing insurance [17]. In these cases, families have the ability to avoid existing systems of regulation, like insurance pre-approvals, to obtain care directly from a provider for a cash payment. Should this trend continue to grow, it may increase the amount of innovative, "last-ditch" care sought by patients and their families, which may be problematic from the standpoint of regulation.

Industry

Lastly, industry funding of neurosurgical innovation is an increasingly problematic area that can result in serious conflicts of interest (discussed at length elsewhere in this book) [6, 14–16, 18]. Neurosurgeons who have novel ideas for device development or operating room technology often approach or are approached by representatives from biomedical device companies. In turn, these professional relationships often become personal, and can affect a surgeon's willingness to use technology from other companies, or compromise their ability to provide objective guidance to their patients in the clinic [14].

Medical device companies aim to improve patient outcomes, just as physicians do. In the end, however, they are also sustained by profits and, much like the public, hope to see a return on investment. These conflicts of interest can lead to a bias toward reporting positive results, overuse of technology that may not be clinically appropriate, and higher costs for patients undergoing procedures [19]. Device manufacturers are constantly under pressure to innovate so that they can stay ahead of the technological curve and outpace their competitors. This financial pressure can then be passed on to neurosurgical consultants in the form of stricter deadlines or looser controls over innovative technology. Additionally, direct-to-consumer advertising by device or pharmaceutical manufacturing companies can further cloud the picture, by coercing untrained patients into the mind-set that a certain treatment or medication is better than another, thus causing them to go seek it from their doctor directly [20–23].

Advocacy for Specific Treatments

More recently, some patients and patient advocacy organizations have begun lobbying the healthcare industry for expanded access to specific treatments. As the role of the Internet and social media continues to expand access for nonmedically trained patients and their ability to "shop around" for different physicians, public pressure for particular treatments has become an increasingly serious issue.

Consider the example of a patient who needs to undergo spine surgery. If this patient lived in a city with two spine surgeons, one of whom has a website offering "minimally invasive" surgery and one of whom offers a classic, open approach, the patient may opt for the former, regardless of whether or not this is the most appropriate approach. Patients who do not fully understand the implications of particular procedures might be misled into believing the superiority of one over another based on a variety of factors, including direct-to-consumer advertising, discussions they have had with friends and family, or even in their conversations with their doctors, who may be subject to significant conflicts of interest [21–23].

Another concerning example in the context of neurosurgery is the recent surge in stem cell procedures being provided to patients with neurological disease [7, 9]. Despite being far from proven, stem cell procedures are frequently offered to patients, often at high personal cost. These for-profit centers advertise their treatments on the Internet and on television, attracting desperate patients with debilitating symptoms. They then provide non-approved "treatments" that often have serious adverse effects [8]. Yet because the public has been deceived into believing that these procedures are beneficial, they frequently seek out clinics that will provide them and pay willingly out of pocket to receive this type of care. This in turn may incentivize neurosurgeons to begin providing such services, because doing so can help expand their referral network and increase their personal financial gain.

Even when specific treatments are not being advocated for directly, patients and patient advocacy organizations still have the ability to influence referral patterns and thereby influence clinical practice. Consider a disease-specific organization like the Cushing's Support and Research Foundation (CSRF), which seeks to increase awareness in the medical community of Cushing's disease and to be a resource for patients and families affected by this particular disease [24]. In addition to providing patients a website with information about new treatments, the CSRF also provides newsletters with educational materials and an online database of members. Although smaller in reach than federal agencies like the NIH, disease-specific organizations like CSRF can provide some funding for research projects or publications, which often focus on public awareness [24]. If a new treatment were to be made available for Cushing's disease, the CSRF would be well-equipped to distribute that information to their members and therefore shares in the responsibility of all parties to provide ethical discussion of these issues.

These patient advocacy organizations can be problematic, however, particularly when they receive funding from industry sponsors [25, 26]. A recent study of 439 randomly selected patient advocacy organizations demonstrated that 67.3% self-reported receiving industry funding [26]. Although most of these organizations received relatively modest funding from device manufacturers, some reported significant industry support. As such, these findings raise concerns over the independence of organizations that are advocating on behalf of patients and may cloud their ability to identify innovative treatments that are safe and beneficial for the patients they aim to help.

Ethical Considerations

The ethical obligations affecting surgeons in general, and neurosurgeons specifically, have been well-described elsewhere [5, 9, 13, 14, 27]. In brief, neurosurgeons are obligated to provide care for their patients that is beneficial to their health and well-being, while they are simultaneously obligated to do no harm—to their patient or to society at large. They must also provide care that the patient chooses willingly and without coercion, that must be provided in a just manner to patients equally and without concern for race, gender, or ability to pay. These four principle concerns of ethical care—beneficence, non-maleficence, autonomy, and justice—can be applied to the public pressures for innovation experienced by neurosurgeons, as well.

Beneficence and Non-maleficence

Although patients and their advocates seek to improve healthcare through the afore-mentioned methods of exerting public pressure, they have not completed medical training and therefore are not fully equipped to distinguish innovation that truly advances healthcare and provides only the deception of progress. As such, care must be taken to compare and contrast at least two types of patient advocacy campaigns. First are those that are based in science and evidence, seeking to advance public awareness and funding for a disease process and treatment plan that has been borne out by the evidence. These campaigns, such as the work of the Jimmy Fund or the CSRF, are beneficial in that they can improve both access to and quality of care for a particular disease. On the other hand are those, like the online communities of stem cell tourism, that are on the "fringe"—either not borne out by evidence from well-designed studies or deliberately misleading to the consumer for the sake of financial gain [7–9].

Danger can arise when patients and their families are able to seek care outside of standard practice, whether through the use of crowdfunding or by identifying neurosurgeons unconcerned with the traditions of ethical practice. In these cases, access to innovative care can be significantly expanded, possibly at risk to the patient. Public pressure for innovation should seek to expand access to innovative care that is funded and researched in a transparent manner and that is responsibly investigated with the safety of patients at the forefront.

Autonomy

The presence of significant conflicts of interest for neurosurgeons threatens provision of care that meets the ethical criteria for autonomy. As the section on funding above highlights, the possibility that patients are coerced into care that is inappropriate, low value, or overpriced can be heightened when influence from industry funding or other conflicts of interest is present. Additionally, when patients advocate specifically for certain types of care, like stem cell procedures, and are willing to pay out of pocket for it, they may influence the field as a whole toward providing that type of service. Often, these patients are being coerced indirectly, through direct-to-consumer device or pharmaceutical marketing or through misleading advertising by neurosurgeons themselves. Conflicts of interest identified above, such as those experienced by patient advocacy organizations, further cloud the ability of a patient to make a truly autonomous decision.

True autonomy, of course, allows the patient to choose the care that they want. This conflict between autonomy for patients seeking innovative care and the obligation of responsible surgeons to restrict that right is a major component of the literature on surgical innovation and is difficult to resolve without compromise [5, 28]. While autonomy ought to be respected, it is the neurosurgeon's responsibility to provide care which, if chosen freely by the patient, is provided ethically. Some neurosurgeons rightly point out that even if they provide care that meets these ethical criteria, there are others who will not, and patients will find their way there. Nevertheless, changes in the trends of clinical behavior can only happen when some neurosurgeons speak out against care that is low value, not evidence-based, or dangerous to the patient. In their professional relationships, neurosurgeons should be willing to discuss openly with patients their concerns about such treatments.

Justice

Lastly, neurosurgeons must be sure to provide care that meets the criteria of distributive justice [5, 27]. Innovative treatments are by definition new and frequently difficult to provide en masse. This is not to say that all innovative treatments therefore do not meet the criteria of distributive justice. Rather, it is important for neurosurgeons providing innovative treatments to identify and address potential conflicts of interest or avoid them as much as possible. For example, in situations where innovation is funded by a public advocacy group or an individual philanthropist, care should be taken to allow for absolute clinical freedom on behalf of the treating neurosurgeon, so that the priorities of the funder cannot be dictated directly through the care provided. This will prevent patients with financial means from dictating the care received by those close to them and, by extension, the care received by those without financial means.

Conclusion

Public pressure for innovation plays a key role in establishing the goals of neurosurgical research and clinical practice. Public, philanthropic, and industry funding of research and innovation each has direct and differing effects on the advances achieved in neurosurgical practice, while public advocacy and pressure for specific treatments can shift clinical practice patterns. These varied forms of public pressure for innovation should meet the criteria of beneficence, non-maleficence, autonomy, and justice, just as the care provided by neurosurgeons should. As a whole, public pressure for innovation must be addressed directly so that neurosurgeons can respond directly to the concerns of the public to achieve outcomes satisfying for both parties.

References

1. Buhles WC. Compassionate use: a story of ethics and science in the development of a new drug. Perspect Biol Med. 2011;54(3):304–15.
2. Mukherjee S. The emperor of all maladies: a biography of cancer. New York: Scribner; 2010.
3. Brinker N, Braun S. The Susan G. Komen Breast Cancer Foundation. Breast Dis. 1998;10(5–6):23–8.
4. Maza J. Patient advocacy profile: Susan G. Komen Breast Cancer Foundation. Clin Adv Hematol Oncol H&O. 2004;2(2):129–30.
5. Meyerson D. Is there a right to access innovative surgery? Bioethics. 2015;29(5):342–52.
6. McKneally MF, Daar AS. Introducing new technologies: protecting subjects of surgical innovation and research. World J Surg. 2003;27(8):930–4; discussion 934–5.
7. Bowman M, Racke M, Kissel J, Imitola J. Responsibilities of health care professionals in counseling and educating patients with incurable neurological diseases regarding "stem cell tourism": caveat emptor. JAMA Neurol. 2015;72(11):1342–5.
8. Berkowitz AL, Miller MB, Mir SA, et al. Glioproliferative lesion of the spinal cord as a complication of "stem-cell tourism". N Engl J Med. 2016;375(2):196–8.
9. Cote DJ, Bredenoord AL, Smith TR, et al. Ethical clinical translation of stem cell interventions for neurologic disease. Neurology. 2017;88(3):322–8.
10. Campbell EG. The future of research funding in academic medicine. N Engl J Med. 2009;360(15):1482–3.
11. Lauer MS, Nakamura R. Reviewing peer review at the NIH. N Engl J Med. 2015;373(20):1893–5.
12. Hudson KL, Collins FS. The 21st century cures act—a view from the NIH. N Engl J Med. 2017;376(2):111–3.
13. Cote DJ, Balak N, Brennum J, et al. Ethical difficulties in the innovative surgical treatment of patients with recurrent glioblastoma multiforme. J Neurosurg. 2017;126(6):2045–50.

14. DiPaola CP, Dea N, Dvorak MF, Lee RS, Hartig D, Fisher CG. Surgeon-industry conflict of interest: survey of opinions regarding industry-sponsored educational events and surgeon teaching: clinical article. J Neurosurg Spine. 2014;20(3):313–21.
15. Janssen SJ, Bredenoord AL, Dhert W, de Kleuver M, Oner FC, Verlaan JJ. Potential conflicts of interest of editorial board members from five leading spine journals. PLoS One. 2015;10(6):e0127362.
16. Hollak CE, Biegstraaten M, Baumgartner MR, et al. Position statement on the role of healthcare professionals, patient organizations and industry in European Reference Networks. Orphanet J Rare Dis. 2016;11:7.
17. Driscoll B. Frankie-Rose's family raise 175,000 in just one week to fund overseas cancer treatment. 2014. http://www.huffingtonpost.co.uk/2014/11/17/frankie-rose-lea-raise-money-cancer-treatment-proton-beam_n_6170624.html. Accessed 2 Mar 2017.
18. Fisher CG, DiPaola CP, Noonan VK, Bailey C, Dvorak MF. Physician-industry conflict of interest: public opinion regarding industry-sponsored research. J Neurosurg Spine. 2012;17(1):1–10.
19. Bailey CS, Fehlings MG, Rampersaud YR, Hall H, Wai EK, Fisher CG. Industry and evidence-based medicine: believable or conflicted? A systematic review of the surgical literature. Can J Surg. 2011;54(5):321–6.
20. Hurst DJ. Restoring a reputation: invoking the UNESCO Universal Declaration on Bioethics and Human Rights to bear on pharmaceutical pricing. Med Health Care Philos. 2017;20(1):105–17.
21. Berndt ER. To inform or persuade? Direct-to-consumer advertising of prescription drugs. N Engl J Med. 2005;352(4):325–8.
22. Donohue JM, Cevasco M, Rosenthal MB. A decade of direct-to-consumer advertising of prescription drugs. N Engl J Med. 2007;357(7):673–81.
23. Greene JA, Watkins ES. The vernacular of risk—rethinking direct-to-consumer advertising of pharmaceuticals. N Engl J Med. 2015;373(12):1087–9.
24. Cushing's Support and Research Foundation. https://csrf.net/. Accessed 24 Feb 2017.
25. Rose SL. Patient advocacy organizations: institutional conflicts of interest, trust, and trustworthiness. J Law Med Ethics. 2013;41(3):680–7.
26. Rose SL, Highland J, Karafa MT, Joffe S. Patient advocacy organizations, industry funding, and conflicts of interest. JAMA Intern Med. 2017;177(3):344–50.
27. Emanuel EJ, Wendler D, Grady C. What makes clinical research ethical? JAMA. 2000;283(20):2701–11.
28. Meyerson D. Innovative surgery and the precautionary principle. J Med Philos. 2013;38(6):605–24.

David J. Cote

Introduction

Innovative surgical treatments for patients with terminal illnesses represent a unique ethical challenge. Unlike established therapies, innovative treatments are, by definition, unproven. They often pose serious risks without promise of a therapeutic effect. Additionally, patients with terminal illnesses themselves are often considered a uniquely vulnerable population. They may be more likely to accept large risks for just a sliver of a chance at success. As such, some ethicists have asserted stronger rights to protection for these patients, and certainly the use of unproven therapies would be considered an area fertile for ethical discussion.

As an example, we might consider the case of a patient with recurrent glioblastoma multiforme (GBM). Despite years of medical and surgical innovation, outcomes for this disease are generally so poor that many patients opt for palliative care at the time of recurrence over aggressive intervention [1–3]. As a result of this poor prognosis, however, surgical innovation in the care of these patients is robust, with current investigations examining the use of fluorescent agents, intratumoral infusion of oncolytic viruses, stem cell transplants, implantable chemotherapeutic agents, and other surgical adjuncts [4–7].

It was in this innovative milieu that in 2013, a neurosurgeon practicing in California purposely inoculated bacteria into the tumor resection cavity of several patients he had operated on for recurrent GBM [8, 9]. He did so based on scattered

Based on: Cote DJ, Balak N, Brennum J, Holsgrove DT, Kitchen N, Kolenda H, Moojen WA, Schaller K, Robe PA, Mathiesen T, Broekman ML. Ethical difficulties in the innovative surgical treatment of patients with recurrent glioblastoma multiforme. J Neurosurg. 2017 Jun;126(6):2045–2050.

D. J. Cote (✉)
Department of Neurosurgery, Computational Neurosciences Outcomes Center (CNOC), Brigham and Women's Hospital, Harvard Medical School, Boston, MA, USA
e-mail: david_cote@hms.harvard.edu

M. L. D. Broekman (ed.), *Ethics of Innovation in Neurosurgery*,
https://doi.org/10.1007/978-3-030-05502-8_10

case reports that showed prolonged survival in patients who developed postoperative infections after resection of central nervous system malignancies [10–13]. These reports suggested that bacterial infection in patients treated for GBM could provoke an immune response in the area of the tumor that may lead to improved survival. His hope was that in provoking an infection, he might paradoxically improve survival.

This case, and the example of innovation for recurrent GBM as a whole, provides an example of the ethical difficulties that may arise when patients who are suffering from a terminal illness approach the end of life. Many of these patients seek innovative surgical procedures or innovative treatments provided by a surgeon that are far from proven. In these cases, a conflict can arise between the duty of surgeons to provide ethically correct care and the fundamental bioethical principle of patient autonomy. Do patients with terminal illnesses have a fundamental right to access innovative, unproven surgical treatments, even if these treatments may pose risk of serious injury?

The current literature on this issue centers around two opposing arguments: on the one hand, that patients nearing the end of life have a right to innovative treatment under the auspices of patient autonomy and compassionate use and, on the other, that access to innovative treatment near the end of life frequently subverts regulation, risks undermining knowledge-generation structures, and poses a serious ethical risk. In this chapter, we will evaluate these arguments in the context of surgical innovation, which is fundamentally different from medical innovation. Much of the discussion is based on a previously published paper, written with members of two international committees on ethics in neurosurgery: the European Association of Neurosurgical Societies (EANS) Ethico-Legal Committee and the World Federation of Neurosurgical Societies (WFNS) Ethics and Medico-Legal Affairs Committee [14]. We will end by evaluating the case of patients with recurrent GBM, to illustrate the protections that should be provided and rights that should be extended to patients with terminal illness seeking surgical innovation.

Surgical Innovation

Discussion of the ethics of surgical innovation first requires sufficient background on the broad realities of surgical innovation in clinical practice and an understanding of the ethical issues at hand. Surgical innovation has historically received less focus in ethical discussions than medical innovation, largely because of the "exceptional" status of surgery, which distinguishes both regulation and innovation of surgical procedures from that of medical treatments [15–17].

Whereas medical innovation is traditionally restricted to well-regulated clinical trials, surgical innovation is part of the daily life of surgery [15]. Each surgical case is slightly different from the one preceding it, and two surgeons may approach the same problem from completely different perspectives. Slight modifications of surgical procedures frequently produce distinct procedures, to the point that these could possibly be called new operations altogether. In nearly all cases, these innovations are not subject to the traditional regulatory structures of a randomized, controlled clinical trial (RCT), largely because RCTs are often impractical in the case of surgical innovation.

Unlike surgical procedures, determining the efficacy of a medical treatment is largely accomplished through RCTs, which depend on a rigid structure that first evaluates the safety of the drug, followed later by an evaluation of its efficacy. In most cases, the effects of a new drug are compared to the current standard of care or a placebo. Physicians that are part of the care team are often blinded to the patient's arm in the trial. In surgical innovation, this design is nearly impossible, but other trial designs are being increasingly recognized as legitimate [16]. To carry out a "Phase I" trial of a surgical innovation, one would need to find a population of volunteers willing to undergo a surgical procedure from which they might not benefit, to demonstrate its relative safety. Further, comparing innovations to the standard of care or to placebo in an ethical manner requires equipoise—the reasonable belief that patients in neither treatment group have a significant advantage over the other—a circumstance that is extremely rare in surgical innovation [18]. Many writers have also questioned the difficult ethical requirements of so-called sham surgery placebos, which are not true placebos but rather are surgical procedures that do not carry out the intended surgical operation, and blinding surgeons to their patient's treatment is nearly always impossible [19–22]. Nevertheless, there are massive potential rewards to surgical innovation, so it is imperative that future regulation does not overly burden the practice to the extent that it risks stifling growth [17].

Although surgical innovation is ubiquitous, there are clearly different types, and defining what is and what is not appropriate has been historically difficult [15, 16, 23–25]. One of the major difficulties in assessing innovative therapies of any kind is defining exactly what types of procedures require assessment. Most surgical procedures exist along a spectrum of innovation: some have been performed dozens or hundreds of times and are undergoing only small modification; others are entirely new and radically different from the standard of care. The respective views of neurosurgeons and ethicists on these procedures differ based on personal experience, and excellent surgeons often disagree on what should or should not be considered innovative. Currently, many investigators are using surveys and other qualitative research methods to define innovation, based on the input of practicing members of various fields. This, "self-regulated" definition of innovation is useful to distinguish between adjacent areas on the spectrum of innovation: minor modifications to a procedure or the use of slightly different equipment is part of the day-to-day reality of surgery, but radical changes to surgical procedures, or new procedures entirely, are generally not [25]. In the context of the ethical discussion presented here, we will largely be considering innovation in the latter categories, which seeks to shift the treatment paradigm through radical innovation.

Autonomy and the Right to Access

It is clear from the above discussion that surgical innovation differs significantly from medical innovation and that the vast ethical discussion on medical innovation and compassionate use cannot necessarily be applied directly to surgical innovation. Already, many writers have discussed the relative difficulties for meeting the high ethical and regulatory standards for medical treatments in a surgical setting, where innovation is constantly occurring in small and incremental ways [16, 23–26]. Moving from these arguments, many have made the claim that access to surgical innovation aligns with patient autonomy and should be permitted in certain circumstances.

Autonomy is a fundamental principle of biomedical ethics that maintains the right of all patients to make decisions over their own care and treatment, free from interference by others [18, 24, 27]. Philosophically based on the human right to self-determination and now legally codified in many countries, autonomy has become an essential part of modern medical practice. Autonomy is the ethical and philosophical justification for both shared decision-making and informed consent and permits patients to have significant control over their treatment course [2, 28].

Autonomy extends even to extreme cases. Ethicists have repeatedly reaffirmed the right of a patient to refuse treatment, even in cases when it will certainly cause her harm or death [29]. This principle has typically been referred to as the "right to die." In the same way, many ethicists have argued that patients nearing the end of life should have a "right to life," which would allow them to choose high-risk, innovative, or unproven treatment strategies as a final effort at extending life [23, 24].

Although the existing literature has focused mainly on medical innovation, some bioethicists have also argued that these rights should extend to access to innovative surgical procedures, when such procedures may extend life or palliate symptoms [28, 30]. For recurrent GBM, these innovative surgical procedures might include, for example, the use of 5-aminolevulinic acid (5-ALA) for fluorescence and photodynamic therapy, a procedure for which evidence is recently mounting [31–34]. While such innovations have been reported to improve outcomes, few have been subjected to the high standard of true RCTs [34]. The informed consent and oversight processes that currently exist allow patients to participate in clinical trials based on their right to autonomy, even at some risk to themselves, and some have argued that this right to assume reasonable risk should be extended to surgical innovation as well [15, 16, 24].

Restraints on the Right to Access

While autonomy is a fundamental human right, many bioethicists have acknowledged that few rights are absolute and that the right to access innovative treatments near the end of life is one that can justifiably be infringed [24, 30, 35–39]. Autonomy, though crucial to biomedical ethics, is just one of many fundamental rights that must be weighed when deciding the appropriateness of an intervention, and other factors, like beneficence and non-maleficence, must also be considered [18, 40].

Allowing access to innovative treatments near the end of life may subvert important knowledge-generation structures, which are closely related to but distinct from regulatory structures [35, 37, 39, 41, 42]. Allowing access to surgical innovations at the end of life outside the structures of these trials may exclude patients from participation in existing RCTs or cloud the results of well-designed studies, thus harming future patients who could benefit from a complete and rigorous analysis. Allowing access to innovative surgery could result in fewer patients participating in RCTs, thereby depriving the public of essential scientific knowledge that may have gone on to benefit future patients [39]. As previously discussed, conducting surgical RCTs for recurrent GBM is already difficult. Policies that allow easier access to surgical innovation outside of RCTs may make conducting such studies nearly impossible.

The high costs of innovative treatments must also be considered. Given the increased risk of many of these procedures, the use of new technologies, and the frequently unknown results, the financial costs of innovative procedures tend to be much higher than those that are considered standard of care [23, 37, 43–46]. Additionally, many typical insurance plans do not cover the costs of innovative treatments. In lieu of typical coverage, who might bear these costs, and how might these costs be reasonably distributed among members of society who require or seek innovative treatment? For now, the lack of coverage leaves many innovative treatments available only to the richest in society, which can exacerbate existing gaps in healthcare quality across social classes [23, 24, 47, 48]. Limiting these disparities is a noble goal that some believe may require temporarily limiting access to innovative treatments, until access can be expanded so that innovative treatments can be made more widely available.

Many also argue that physicians have a fundamental obligation to protect their patients from unethical treatments. This argument can most aptly be summarized by stating that a physician's obligation to her patients should be based on an assessment of the current evidence base and that the use of innovative yet unproven treatments outside of investigative structures may be unethical because such a treatment could be harmful to the patient or to others who might benefit from any generalizable knowledge that is being sacrificed [24, 30, 39, 49]. Innovations outside traditional regulatory structures may truly be dangerous, and it is the duty of physicians to steer their patients away from situations in which they may be vulnerable to false promises or harmful treatments.

Lastly, from an ethical perspective, many who argue that the right to innovative treatment can justifiably be infringed point out that whereas the so-called right to die is a negative right, the right to innovative treatment would be a positive right [24]. Positive rights, unlike negative rights, yield power to the right holder that she can wield over others. Particularly from the perspective of a medical professional, asserting that a patient should have the right to receive surgical treatment may infringe on the rights of surgeons to refuse performing such operations [24, 39].

Ethical Evaluation of Surgical Innovations

If one respects patient autonomy, there must be certain circumstances in which surgical innovation can be carried out ethically and in which patients can consent to innovative procedures. Based on this fundamental ethical principle, the right to innovative treatment should be presumed but should be restrained in certain circumstances. Below are four different conditions that should be met for surgical innovation to be considered ethical, each of which also highlights the ways in which surgical innovation differs from medical innovation on ethical grounds. If any of these conditions are not met and surgical innovation is deemed unethical, a patient's right to innovative surgical treatments can be justifiably infringed.

Informed Consent and Vulnerable Patients

Informed consent is a fundamental principle of ethical clinical research that depends significantly on the right of a patient to autonomy [2, 15, 18, 40, 50–53]. When undergoing an innovative surgical procedure, informed consent requires a higher standard than in the course of normal clinical care, due to the lower quality of evidence of the intervention and the unique nature of experimentation in the surgical patient [15].

First, surgeons carrying out the innovative procedure must disclose to the patient the exact nature of the innovation, specifically stating that the strategy being used is not part of the standard treatment for the patient's disease [54]. The treatment team must also disclose alternative treatment options to the patient, including the relative risks and benefits of each option [18].

For patients with terminal illnesses, informed consent takes on considerably more importance due to the increased vulnerability of the patient population [50, 51, 55, 56]. This vulnerability is partially because they may not be able to protect their own interests and because they have not voluntarily chosen their status of having a terminal illness [57]. Although there has been some debate about this issue, patients with terminal illnesses are generally considered to be more willing to assume risks and choose treatments with lower chances of potential benefits [55, 58]. In these cases, the informed consent process must be considerably more formal and requires a lengthy, face-to-face discussion, with attention paid specifically to the unique situation of the terminally ill patient [2, 18, 40, 58].

Additionally, patients should have their capacity to make medical decisions evaluated and confirmed. Near the end of life, patients must meet at least four criteria that determine capacity: (1) that the patient is able to communicate a choice, (2) that they understand the information relevant to that choice, (3) that they appreciate the situation and its consequences, and (4) that they can reason through the options being presented to them [50]. Patients who have neurocognitive deficits or psychiatric illnesses, including depression, may not be deemed to have the capacity to make decisions about their care; such patients cannot ethically receive innovative surgical treatments [15].

Oversight and Regulation

Because surgical innovation is not generally considered human subject research, it often does not fall under the regulatory structure of an institutional review board

(IRB) [30, 37, 42, 59]. Nevertheless, surgical innovation does fall under the auspices of the IRB if certain conditions are met [15]. These include cases where a surgeon is testing a formal hypothesis and attempting to produce generalizable knowledge, or if a surgeon plans to perform an innovative procedure repeatedly, on multiple patients [15, 16, 22–24]. Many surgical innovations occur outside these conditions, however, and these innovations should not go without any oversight or regulation.

Previous writers have described the formation of a surgical innovation committee (SIC), a committee similar to an IRB tasked with reviewing surgical innovations that do not fall under the auspices of the IRB itself and are not part of the innovative nature of surgery [15]. These cases include operations that differ significantly from the standard of care or that have not been rigorously studied or previously reported [15]. The SIC would independently monitor surgical innovations submitted for review, assess the relevant risks and benefits in the context of the current standard of care, ensure informed consent and a reasonable chance of success, and approve or deny submissions on a case-by-case basis. Postoperatively, the SIC would ensure adequate follow-up of patients undergoing innovative procedures and mandate the reporting of outcomes and adverse events. The development of SICs at investigative medical centers nationwide would allow for ethical surgical innovation without damaging the current culture of innovation crucial to the success and future development of surgical specialties.

Efficacy of the Innovative Neurosurgical Treatment

For a surgical innovation to be ethical, there must be a reasonable belief that it could be as successful as or more successful than the current standard of care [15, 18, 24]. This principle depends primarily on the idea of non-maleficence and relates to the previous discussion of equipoise. In this sense, the ethical evaluation of surgical treatments is no different than the ethical evaluation of medical treatments. If a surgeon believed that a patient could benefit from the standard of care, but subjected that patient to an innovative, unproven treatment that did not have a reasonable chance at success, this practitioner would be committing an ethical violation, even if the patient were to provide truly informed consent.

Unfortunately, determining the relative efficacy of a surgical innovation is extremely difficult or even impossible preoperatively, and drawing the line between treatments that are completely unproven and those that may have a reasonable chance at success is difficult. Predicting success depends both on the medical community's knowledge of disease processes and pathophysiology, which is incomplete, and on predictions of the relative merits of the innovation, which would undoubtedly be tenuous in a procedure being carried out for the first time. Thus the evaluation of a surgical innovation's efficacy should not be determined solely by the performing surgeon, but rather should be summarized and presented by that surgeon to an independent SIC or some similar alternative, which could then evaluate the innovative procedure in the context of the current standard of care. Alternatives to formal SICs include peer review by a surgeon's local, national, or international colleagues, a dedicated form of IRB, the input of an external institution, or review by the surgeon in chief [25]. Surgical innovations that place an unreasonable burden on the patient, pose a serious risk of morbidity or mortality, or do not have sufficient evidence in animal models should be deemed unethical until more information is made available.

Harm to Others

A key argument against compassionate use in the context of medical treatments is the subversion of knowledge-generation structures, as previously mentioned [39]. In surgical innovation, this is generally not an issue, because so few surgical procedures are subject to the rigid structure of RCTs in the first place. In fact, surgical procedures are often driven forward by technical case reports of one or a handful of patients, which can sometimes revolutionize surgical care.

As such, harm to others is a less relevant but still important consideration in ethical surgical innovation. Surgical innovations could directly harm the patients involved if the discussion and evaluation of possible risks preoperatively are inadequate. Additionally, patients seeking surgical innovations may harm current knowledge-generation structures (e.g., medical clinical trials for this disease) if these innovations were attempted prior to participation in clinical trials and later resulted in exclusion of patients from participation, or family and loved ones, who play a crucial role in supporting patients through a difficult treatment course. Truly innovative surgical treatments for patients approaching the end of life should generally be carried out only after exhausting other reasonable options.

The Case of Recurrent GBM

Throughout this chapter, we have discussed the many criteria that must be weighed when considering whether an innovative surgical procedure can ethically be offered to a patient nearing the end of life. It is helpful to take these more abstract concepts and apply them directly to a neurosurgical case. Patients with recurrent GBM, as mentioned previously, frequently find themselves in the situation of having exhausted all conventional treatment options. Often, they will have undergone at least one surgical procedure, as well as radiation and chemotherapy. When the tumor recurs, there is currently almost no established treatment guideline for further care; therefore, many patients seek innovative surgical procedures.

These procedures rarely affect clinical trials, because so few clinical trials are in place for recurrent GBM, and patients who do have access to these trials often seek truly innovative surgery only after attempting treatment through these trials. In the presence of adequate discussion of the risks and benefits and truly informed consent, there is little risk of harm to others. Additionally, many of these innovations also have reasonable evidence of efficacy. An example previously mentioned,

5-ALA fluorescence, undoubtedly has grown in popularity as its evidence base similarly grows. Lastly, these treatments are usually (and should always be) carried out in supervised environments, with adequate data collection and oversight by a SIC or committee of colleagues with adequate independence.

Conclusion

As we have discussed previously in work published with the ethics committees of the WFNS and EANS, patients with terminal illnesses should have a right to innovative surgical treatments, as this aligns with the fundamental ethical principle of autonomy [14]. This right is not absolute, however, and reasonable and appropriate measures should be taken to ensure adequate protection of these vulnerable patients. These measures include (1) a high standard of truly informed consent, with attention given specifically to the vulnerability of the patient, the innovative aspects of the procedure, and the capacity of the patient to consent; (2) substantial oversight and regulation of the innovative treatment, preferably in the form of an SIC or, when appropriate, an IRB; (3) adequate evidence that the innovative treatment will be successful, either in the form of animal model studies or prior use of closely related procedures in humans; and (4) no risk of harm to others. If these standards are not met, a patient's right to innovative surgical treatment can be justifiably infringed.

References

1. De Bonis P, Fiorentino A, Anile C, et al. The impact of repeated surgery and adjuvant therapy on survival for patients with recurrent glioblastoma. Clin Neurol Neurosurg. 2013;115(7):883–6.
2. Agar M, Ko DN, Sheehan C, Chapman M, Currow DC. Informed consent in palliative care clinical trials: challenging but possible. J Palliat Med. 2013;16(5):485–91.
3. Ringel F, Pape H, Sabel M, et al. Clinical benefit from resection of recurrent glioblastomas: results of a multicenter study including 503 patients with recurrent glioblastomas undergoing surgical resection. Neuro-Oncology. 2016;18(1):96–104.
4. Craig SE, Wright J, Sloan AE, Brady-Kalnay SM. Fluorescent-guided surgical resection of glioma with targeted molecular imaging agents: a literature review. World Neurosurg. 2016;90:154–63.
5. Wait SD, Prabhu RS, Burri SH, Atkins TG, Asher AL. Polymeric drug delivery for the treatment of glioblastoma. Neuro-Oncology. 2015;17(Suppl 2):ii9–ii23.
6. Cho DY, Lin SZ, Yang WK, et al. Targeting cancer stem cells for treatment of glioblastoma multiforme. Cell Transplant. 2013;22(4):731–9.
7. Cote DJ, Bredenoord AL, Smith TR, et al. Ethical clinical translation of stem cell interventions for neurologic disease. Neurology. 2017;88(3):322–8.
8. Eakin E. Bacteria on the brain. New York: The New Yorker; 2015.
9. Ellor SV, Pagano-Young TA, Avgeropoulos NG. Glioblastoma: background, standard treatment paradigms, and supportive care considerations. J Law Med Ethics. 2014;42(2):171–82.
10. De Bonis P, Albanese A, Lofrese G, et al. Postoperative infection may influence survival in patients with glioblastoma: simply a myth? Neurosurgery. 2011;69(4):864–8; discussion 868–869.
11. Bohman LE, Gallardo J, Hankinson TC, et al. The survival impact of postoperative infection in patients with glioblastoma multiforme. Neurosurgery. 2009;64(5):828–34; discussion 825–834.

12. Bowles AP Jr, Perkins E. Long-term remission of malignant brain tumors after intracranial infection: a report of four cases. Neurosurgery. 1999;44(3):636–42; discussion 633–642.
13. Kapp JP. Microorganisms as antineoplastic agents in CNS tumors. Arch Neurol. 1983;40(10):637–42.
14. Cote DJ, Balak N, Brennum J, et al. Ethical difficulties in the innovative surgical treatment of patients with recurrent glioblastoma multiforme. J Neurosurg. 2017;126(6):2045–50.
15. Biffl WL, Spain DA, Reitsma AM, et al. Responsible development and application of surgical innovations: a position statement of the Society of University Surgeons. J Am Coll Surg. 2008;206(6):1204–9.
16. Barkun JS, Aronson JK, Feldman LS, et al. Evaluation and stages of surgical innovations. Lancet (London, England). 2009;374(9695):1089–96.
17. London AJ. Cutting surgical practices at the joints: individuating and assessing surgical procedures. In: Reitsma AM, Moreno JD, editors. Ethical guidelines for innovative surgery. Hagerstown, MD: University Publishing Group; 2006.
18. Emanuel EJ, Wendler D, Grady C. What makes clinical research ethical? JAMA. 2000;283(20):2701–11.
19. Niemansburg SL, van Delden JJ, Dhert WJ, Bredenoord AL. Reconsidering the ethics of sham interventions in an era of emerging technologies. Surgery. 2015;157(4):801–10.
20. Galpern WR, Corrigan-Curay J, Lang AE, et al. Sham neurosurgical procedures in clinical trials for neurodegenerative diseases: scientific and ethical considerations. Lancet Neurol. 2012;11(7):643–50.
21. Cohen PD, Isaacs T, Willocks P, et al. Sham neurosurgical procedures: the patients' perspective. Lancet Neurol. 2012;11(12):1022.
22. Angelos P. Ethical issues of participant recruitment in surgical clinical trials. Ann Surg Oncol. 2013;20(10):3184–7.
23. Meyerson D. Innovative surgery and the precautionary principle. J Med Philos. 2013;38(6):605–24.
24. Meyerson D. Is there a right to access innovative surgery? Bioethics. 2015;29(5):342–52.
25. Broekman ML, Carriere ME, Bredenoord AL. Surgical innovation: the ethical agenda: a systematic review. Medicine (Baltimore). 2016;95(25):e3790.
26. Tsou A. Ethical considerations when counseling patients about stem cell tourism. Continuum (Minneapolis, Minn.). 2015;21(1 Spinal Cord Disorders):201–5.
27. Sacristan JA, Aguaron A, Avendano-Sola C, et al. Patient involvement in clinical research: why, when, and how. Patient Prefer Adherence. 2016;10:631–40.
28. Walker MJ, Rogers WA, Entwistle V. Ethical justifications for access to unapproved medical interventions: an argument for (limited) patient obligations. Am J Bioeth. 2014;14(11):3–15.
29. van Wijmen MP, Pasman HR, Widdershoven GA, Onwuteaka-Philipsen BD. Continuing or forgoing treatment at the end of life? Preferences of the general public and people with an advance directive. J Med Ethics. 2015;41(8):599–606.
30. Romano LV, Jacobson PD. Patient access to unapproved therapies: the leading edge of medicine and law. J Health Life Sci Law. 2009;2(2):45, 47–72.
31. Eyupoglu IY, Hore N, Merkel A, Buslei R, Buchfelder M, Savaskan N. Supracomplete surgery via dual intraoperative visualization approach (DiVA) prolongs patient survival in glioblastoma. Oncotarget. 2016;7:25755.
32. Hervey-Jumper SL, Berger MS. Maximizing safe resection of low- and high-grade glioma. J Neuro-Oncol. 2016;130:269.
33. Quick-Weller J, Lescher S, Forster MT, Konczalla J, Seifert V, Senft C. Combination of 5-ALA and iMRI in re-resection of recurrent glioblastoma. Br J Neurosurg. 2016;30(3):313–7.
34. Stummer W, Pichlmeier U, Meinel T, Wiestler OD, Zanella F, Reulen HJ. Fluorescence-guided surgery with 5-aminolevulinic acid for resection of malignant glioma: a randomised controlled multicentre phase III trial. Lancet Oncol. 2006;7(5):392–401.
35. Caplan A. Is it sound public policy to let the terminally ill access experimental medical innovations? Am J Bioeth. 2007;7(6):1–3.

36. Caplan A, Bateman-House A. Compassion for each individual's own sake. Am J Bioeth. 2014;14(11):16–7.
37. Caplan AL, Bateman-House A. Should patients in need be given access to experimental drugs? Expert Opin Pharmacother. 2015;16(9):1275–9.
38. Caplan AL, Ray A. The ethical challenges of compassionate use. JAMA. 2016;315(10):979–80.
39. Bender S, Flicker L, Rhodes R. Access for the terminally ill to experimental medical innovations: a three-pronged threat. Am J Bioeth. 2007;7(10):3–6.
40. Beecher HK. Ethics and clinical research. N Engl J Med. 1966;274(24):1354–60.
41. Nycum G, Reid L. The harm-benefit tradeoff in "bad deal" trials. Kennedy Inst Ethics J. 2007;17(4):321–50.
42. Hyun I. Allowing innovative stem cell-based therapies outside of clinical trials: ethical and policy challenges. J Law Med Ethics. 2010;38(2):277–85.
43. Gusmano MK. Is it reasonable to deny older patients treatment for glioblastoma? J Law Med Ethics. 2014;42(2):183–9.
44. Dresser R. "Right to Try" laws: the gap between experts and advocates. Hast Cent Rep. 2015;45(3):9–10.
45. Jacob JA. Questions of safety and fairness raised as right-to-try movement gains steam. JAMA. 2015;314(8):758–60.
46. Caplan A. Medical ethicist Arthur Caplan explains why he opposes 'Right-to-Try' laws. Oncology (Williston Park, N.Y.). 2016;30(1):8.
47. Darrow JJ, Sarpatwari A, Avorn J, Kesselheim AS. Practical, legal, and ethical issues in expanded access to investigational drugs. N Engl J Med. 2015;372(3):279–86.
48. McKneally MF, Daar AS. Introducing new technologies: protecting subjects of surgical innovation and research. World J Surg. 2003;27(8):930–4; discussion 934–935.
49. Buhles WC. Compassionate use: a story of ethics and science in the development of a new drug. Perspect Biol Med. 2011;54(3):304–15.
50. Appelbaum PS. Clinical practice. Assessment of patients' competence to consent to treatment. N Engl J Med. 2007;357(18):1834–40.
51. Grisso T, Appelbaum PS. Assessing competence to consent to treatment. New York: Oxford University Press; 1998.
52. Beecher HK. Consent in clinical experimentation: myth and reality. JAMA. 1966;195(1):34–5.
53. Beecher HK. Some guiding principles for clinical investigation. JAMA. 1966;195(13):1135–6.
54. Lerner BH. Sins of omission—cancer research without informed consent. N Engl J Med. 2004;351(7):628–30.
55. Roy DJ, MacDonald N. Ethical issues in palliative care. In: Doyle D, Hanks GW, MacDonald N, editors. Oxford textbook of palliative medicine. Oxford: Oxford University Press; 1998. p. 97–138.
56. Brody BA. The ethics of biomedical research: an international perspective. New York: Oxford University Press; 1998.
57. Canavero S, Bonicalzi V. Central pain following cord severance for cephalosomatic anastomosis. CNS Neurosci Ther. 2016;22(4):271–4.
58. Casarett DJ, Karlawish JH. Are special ethical guidelines needed for palliative care research? J Pain Symptom Manag. 2000;20(2):130–9.
59. Roth-Cline M, Nelson R. FDA implementation of the expanded access program in the United States. Am J Bioeth. 2014;14(11):17–9.

Ethics Committees, Innovative Surgery, and Organizational Ethics

Joseph P. Castlen and Thomas I. Cochrane

Case Report: Uterine Transplant Funding

After a hospital employee at a large American teaching hospital faced the possibility of needing a hysterectomy, interest was sparked in the formation of a uterine transplant program. Although the employee ended up not needing the procedure, a multidisciplinary working group was formed to explore the possibility of developing a uterine transplant program. This group included transplant surgeons, adult and pediatric gynecologists, fertility physicians, and reproductive surgeons. Since uterine transplantation was a new field (at the time, only 11 transplants had been successfully performed worldwide), insurance companies would not cover the procedure. As such, the group began looking for ways to fund the initial transplant procedures until more support could be obtained.

Without insurance approval and in the absence of any philanthropic funding, the working group approached the hospital administration to request funding for the first few uterine transplants. It was their belief that patients should not have to pay for an innovative treatment with uncertain benefits and, on a related note, that patients should be selected without regard to their ability to pay. In addition, this hospital had a history of funding innovative procedures, including a leading face transplantation program. The multidisciplinary team at this hospital was uniquely

J. P. Castlen
Department of Neurosurgery, Computational Neurosciences Outcomes Center, Brigham and Women's Hospital, Harvard Medical School, Boston, MA, USA

T. I. Cochrane (✉)
Department of Neurology, Brigham and Women's Hospital, and Center for Bioethics, Harvard Medical School, Boston, MA, USA
e-mail: tcochrane@partners.org

© Springer Nature Switzerland AG 2019
M. L. D. Broekman (ed.), *Ethics of Innovation in Neurosurgery*,
https://doi.org/10.1007/978-3-030-05502-8_11

equipped to manage such a novel procedure, and part of the hospital's stated mission was to be a world leader in medical and surgical innovation.

Hospital administration did not initially want to fund the first few procedures, arguing primarily that it was in a worse financial situation than when it had previously funded other innovative transplantation programs. Administrators claimed that if they funded the uterine transplants, it would divert resources from other programs, though it was not immediately clear which specific programs would suffer a loss of resources. Some administrators felt that uterine transplantation was unlike other transplantation, because it would be performed with the goal of allowing a woman to deliver a healthy baby, and (the argument ran) thus, uterine transplantation would not affect a patient's quality of life in the same way as, for instance, a face transplant. Some members of the working group argued that for some women, being able to carry a pregnancy to term affects their quality of life as much as getting a face or hand transplant.

Hospital administrators requested consultation from the hospital ethics committee (EC), to help determine whether the hospital had an ethical obligation to fund the uterine transplant. After discussion at a regularly scheduled EC meeting, the EC convened a meeting of its members and other stakeholders for a formal ethics consultation. The EC, which was comprised of community members, nursing staff, physicians, and administrators, approached the case much like it approached clinical ethics consultations. After discussing the facts of the case, clarifying the ethical question at hand, engaging in a discussion with both sides, and considering the potential options to resolve the question, the committee resolved that the hospital had a moderately strong ethical obligation to fund the first few uterine transplants, but that financial concerns, if they involved serious tradeoffs that would sacrifice other important goals, could outweigh this obligation. The hospital did not in the end fund the uterine transplant program.

Case Analysis

Although this EC was not typically consulted for organizational ethics questions, the members of the committee approached this consultation similarly to clinical ethics consults. Ultimately, the committee gave the qualified recommendation that the hospital should fund the procedures but that it was not obliged to do so at the expense of other important clinical services. It also resolved that patients should not be allowed to pay for the initial procedures, for two reasons: first, it did not seem appropriate to ask patients to pay out of pocket for an as-yet-unproven therapy and, second, the committee thought it was important to avoid creating a system in which only wealthy individuals had access to innovative care. Another benefit of this consultative process was that the committee was able to provide a method whereby administrators and clinicians could carefully deliberate regarding whether the hospital had an obligation to fund the initial transplants.

This consultation was directly related to the funding of an innovative procedure. However, we feel it illustrates how hospital ECs can be used to address organizational ethics questions in general. As medicine becomes more costly, complicated, and difficult to manage, there are bound to be questions of how to regulate and pay for innovative surgical procedures. Pre-existing hospital ECs, which already practice methods of deliberation regarding difficult ethical questions, could be a valuable resource for administrators and caregivers in preventing and resolving ethical questions surrounding innovative surgery.

One of the major questions that the committee considered was whether patients who could benefit from a uterine transplant had a similar level of "medical need" as patients who might benefit from hand or face transplants. Recognizing that this was a question best answered by clinicians with direct experience treating patients with uterine factor infertility, the committee thus dedicated much of its time to soliciting input from the fertility specialists who would be most likely to refer patients for uterine transplantation.

Early in the consultative process, committee members identified a potential problem regarding the consultation. The committee was asked to make recommendations regarding the hospital's obligation to fund an innovative surgical procedure. However, the committee did not have detailed financial information regarding other funding priorities that might have to be sacrificed if the hospital funded the novel transplant program. Furthermore, the committee lacked the expertise to evaluate fiscal considerations in detail. This could have put the committee in an uncomfortable position, as it tried to develop an ethically defensible recommendation regarding a topic which it lacked information and expertise.

In response, the committee made it clear to hospital administrators that the recommendations would have to come in the form of somewhat hypothetical guidelines and that the committee could not make a final and specific recommendation about what the hospital *ought to do*. As a result, the final recommendations came in the following form: "The hospital appears to have a moderately strong ethical obligation to financially support the uterine transplant program. This obligation arises partly from the hospital's stated mission(s) regarding innovative clinical care. An obligation is also generated, based on a fairness principle, by the hospital's recent financial support for other innovative transplant programs. However, the hospital could plausibly argue that its changing financial situation has lessened the strength of this new obligation, relative to its other obligations (for example, to continue to fund already-existing clinical activities)."

One other notable aspect of this case is that the administration was the initiator of the ethics consultation. The consultation itself, however, included input from both the administration and the clinical team. The fact that the administration initiated the process indicates that there was a familiarity, if not a certain level of comfort, with the ethics consultation process and committee itself, likely due to quarterly check-in meetings between the EC and the administration.

This case illustrates how a hospital EC can be used to evaluate ethical issues surrounding innovative treatments while also highlighting some limitations of this approach. Formalizing the process for seeking organizational ethics consultations and increasing awareness that it is available will become necessary as this field expands and organizational ethics consultations become more common.

The Evolution of Hospital Ethics Committees

Originally small groups composed of a majority of physicians, hospital ethics committees (ECs) initially came into existence as a result of court cases in the 1970s [1]. In particular, the Karen Quinlan case in 1975, in which Ms. Quinlan entered a vegetative state and her doctors refused to remove her ventilator at the request of her parents, led to a ruling from the New Jersey Supreme Court recommending that "prognosis committees" be used in place of the courts in these cases [2]. This established an advisory role in hospitals for what became clinical ethics committees.

After a subsequent early 1980s President's Commission recommendation and endorsements from multiple medical professional societies, the role of ECs grew [1, 3]. ECs proliferated in hospitals, increasing in presence from 1% to over 60% [1]. They also became more multidisciplinary as literature emerged advocating for "ideal" committees [3]. By the early 1990s, most American hospitals had formed ECs after a Joint Commission on Accreditation of Healthcare Organizations (JCAHO) requirement that hospitals have a formal process for addressing ethical issues in patient care [4, 5]. Now, ECs have a greater representation of nurses, administrators, and community members, although physician leadership on the committees is still common [3].

Modern ECs have evolved to include on-call and "embedded" clinical ethicists to advise clinicians in cases where a full-committee ethics consultation may not be feasible [6]. Including ethics consultants in a hospital intensive care unit (ICU), for instance, may encourage additional ethics consultations and help make ECs more approachable to providers, patients, and their families [7]. Although the initial objective of ECs was to provide consultations on specific issues arising from patient care, their scope has also expanded in the past couple of decades to include policy development and education [8, 9]. ECs are mostly advisory, although in some cases they are used to adjudicate disagreements [10].

What Is Organizational Ethics?

The ethics of an organization are most simply described as the sum of the individual ethical decisions made by members of an organization. It can also be understood as the set of ethical rules—both explicit and implicit—by which an organization governs itself [11]. This is affected in part by the culture of an organization and in part by its policies. The tone set by the organization's leadership largely shapes the culture of the organization and the decisions made by its employees, while the organization's policies act as a blueprint for certain behaviors within an organization [11].

As an area of study, organizational ethics in medicine is relatively new. The term was first used in medical literature in 1991 and was only used sporadically until the turn of the century [12]. In the past decade, the term has come into more common use, and the study of organizational ethics as a field has expanded to include the myriad ethical issues which fall under its umbrella. Among these topics are power dynamics in the operating room, coordination of appropriate transitional care, and accountability in quality improvement efforts [13–15].

Society holds physicians to high standards of ethical conduct due to their control over patient well-being and the monopoly that they hold over their own practice. As such, there is tight self-regulation in place, including the existence of ECs, to help prevent major breaches of accepted ethical practices. Although physicians have influence over clinical ethics questions and some organizational ethics questions (e.g., power dynamics in the operating room), other questions are primarily addressed by administrators (e.g., accountability in quality improvement). ECs have developed over decades to provide guidance and counseling for healthcare providers facing tough ethical questions, but no such body exists for administrators with organizational ethical quandaries. Since all decisions in hospitals ultimately impact the patients, it makes sense to include ECs on these deliberations, whether they are clinical or organizational in nature.

Practically, it makes sense to assign responsibilities for organizational ethics consultations to existing hospital ECs. The infrastructure is already there, the committees are already formed, and committee members already have experience with ethical deliberation. As illustrated in the uterine transplant case, taking a similar approach to clinical and organizational ethics consultations can be effective.

A stronger organizational ethic will ultimately contribute to a stronger medical ethic within a hospital. One area of organizational ethics, surgical innovation, has in the past decade come under increased scrutiny in regard to its regulation and the ethics surrounding the various stages of innovation. Hospital ECs may be able to provide an advisory role in the questions that occur during the development, implementation, and regulation of these procedures.

Ethics Committees and Surgical Innovation

Given their existing role in policy formation, ECs seem naturally suited to an organizational ethics role. By building relationships with hospital leaders, they can help promote a positive culture throughout the organization whereby ethics is viewed as an advisory tool rather than a potential obstacle to efficiency [16].

Inherent in such a role, however, is the risk that the EC is used as a procedural tool of the administration to obtain an official "blessing" for potentially controversial decisions, or that ECs may have an unconscious bias in favor of the priorities of the administration, if their relationships become too close. But this is already a risk in clinical consultation—it is often the case that clinicians requesting an ethics consult may have a certain preferred outcome in mind, and it is unavoidable that ethics consultants will have longstanding relationships with those same clinicians. This does not (and in our view, should not) make it impossible to provide unbiased ethical guidance.

We do, however, believe that some procedural barriers should exist between a hospital EC and the administration which it serves. It seems obvious that EC members who also work in hospital administration should recuse themselves as primary consultants in organizational ethics consultations. We believe ECs should carefully consider what role, if any, hospital administrators should have in the EC and should err on the side of caution. For example, when administrative input is needed at an EC meeting (e.g., when discussing hospital ethics policies), representatives of the administration are best viewed as serving an advisory role to the EC, rather than being viewed as full members of the committee.

It seems unavoidable that on some level, ECs will have to answer to hospital administrative bodies [3]. Since this is largely a result of the way in which committees are funded and formed, and since it may be unavoidable due to the logistical problems with having an "external" EC weigh in on internal matters, the independence of ECs should be emphasized in mission statements, and EC members should have a means to address any potential conflicts of interest without fear of reprisal [17, 18]. This does not mean, however, that ECs and administrators cannot nurture a collegial and collaborative relationship with the shared goal of achieving the most ethically defensible outcomes. We think that internal ECs can provide ethical guidance, as long as institutional roles are clearly defined, and care is taken to constitute the EC membership and procedures in a way that avoids excessive bias.

At the institution described in the case, for example, EC leaders attempt to nurture a collaborative relationship with the administration while taking care to maintain a proper level of independence and detachment. Leaders of the EC meet quarterly with representatives of the hospital administration, to discuss the activities of the EC and also to solicit input and ideas from the administration regarding ethical questions or dilemmas faced by the hospital leadership. It seems that such a simple quarterly check-in meeting is unlikely to influence committee members any more than, for instance, being embedded in an ICU would influence clinical ethics consult members to side with clinicians. In fact, regular check-ins may promote increased ethics consultations by the administration for organizational ethics questions. By having an established relationship with the EC, the option of an ethics consultation will naturally be more prevalent in the minds of administrators facing potential ethical dilemmas. The increased education possible through these meetings may also spur additional consultations by bringing issues into the light which may not have previously been considered "ethics problems."

Given their very nature, innovative treatments require special attention to ethical considerations, including the institutional role in promoting and funding such innovation [19]. Innovative surgical procedures, meanwhile, remain relatively unregulated when compared to formal research programs, which are regulated by IRBs.

One possible method for managing the ethical risk and tension involved in surgical innovation would be to insist that most surgical innovation be studied using randomized controlled trials. Recognizing this, in 2009, members of the IDEAL Collaboration suggested a framework for a more standardized and regulated approach to surgical innovation [20–22]. However, this framework might not be suitable for every type of surgical innovation, and too strict adherence to the framework could potentially result in a hindrance to surgical progress. Many surgeons feel that RCTs are not well-suited to innovative surgical procedures, and regulation remains a challenge [23].

Even as individual studies and the field itself move toward a more transparent and regulated model, there will likely be some institutional involvement in regulation, and ECs should be involved in addressing localized ethical issues. As regulatory bodies are formed for surgical innovation, there also may be a need for organizational ethics consults if issues of compliance, funding, or implementation arise in an increasingly complex regulatory environment.

Conclusion

Although it is a new field of study, organizational ethics is going to play an increasing role in surgical innovation. Existing hospital ECs can help to advise hospital leadership on the most ethically defensible directions to take during the development, implementation, and regulation of these novel procedures. Some training in organizational ethics may be appropriate for EC members, most of which are primarily familiar with clinical ethics issues. The independence of these committees is paramount, lest public perception shifts to a cynical view of hospital administration and distrust of hospital ECs. Ultimately, it will be up to hospital leadership to embrace hospital ECs as an advisory tool to resolve potential conflicts rather than treating them as obstructions to implementing new policies.

References

1. Aulisio MP. Why did hospital ethics committees emerge in the US? AMA J Ethics. 2016;18(5):546–53.
2. Fine RL. From Quinlan to Schiavo: medical, ethical, and legal issues in severe brain injury. Proc (Bayl Univ Med Cent). 2005;18(4):303–10.
3. Courtwright A, Jurchak M. The evolution of American Hospital Ethics Committees: a systematic review. J Clin Ethics. 2016;27(4):322–40.
4. McGee G, Caplan AL, Spanogle JP, Asch DA. A national study of ethics committees. Am J Bioeth. 2001;1(4):60–4.
5. Joint Commission Accreditation Manual for Hospitals. Hosp Food Nutr Focus. 1992;9(3):1, 3–5.
6. Bruce CR, Majumder MA, Stephens A, Malek J, McGuire A. Cultivating administrative support for a Clinical Ethics Consultation Service. J Clin Ethics. 2016;27(4):341–51.
7. Chen YY, Chu TS, Kao YH, Tsai PR, Huang TS, Ko WJ. To evaluate the effectiveness of health care ethics consultation based on the goals of health care ethics consultation: a prospective cohort study with randomization. BMC Med Ethics. 2014;15:1.
8. Lemiengre J, Dierckx de Casterlé B, Schotsmans P, Gastmans C. Written institutional ethics policies on euthanasia: an empirical-based organizational-ethical framework. Med Health Care Philos. 2014;17(2):215–28.
9. Magelssen M, Pedersen R, Førde R. Novel paths to relevance: how clinical ethics committees promote ethical reflection. HEC Forum. 2016;28(3):205–16.
10. Aulisio MP, Arnold RM. Role of the ethics committee: helping to address value conflicts or uncertainties. Chest. 2008;134(2):417–24.
11. Brown M. Ethics in organizations. Issues Ethics. 1989;2(1). https://legacy.scu.edu/ethics/publications/iie/v2n1/homepage.html.

12. Blustein J. Organizational ethics and health care providers. APA Newsl Philos Med. 1991;90(2):6–10.
13. Pinney S, Ho A. The Puck Stops Here: taking organizational accountability seriously. Healthc Q. 2016;18(4):20–4.
14. Berlinger N, Dietz E. Time-out: the professional and organizational ethics of speaking up in the OR. AMA J Ethics. 2016;18(9):925–32.
15. Naylor M, Berlinger N. Transitional care: a priority for health care organizational ethics. Hast Cent Rep. 2016;46(Suppl 1):S39–42.
16. Sabin JE. How can clinical ethics committees take on organizational ethics? Some practical suggestions. J Clin Ethics. 2016;27(2):111–6.
17. Wolf JS. Point and counterpoint. Should HECs report to the medical staff rather than to the administration, board of trustees, or other administrative office? Yes. HEC Forum. 1993;5(2):115–7.
18. Wolf JS, deBlois J. Point and counterpoint: should HECs report to the medical staff rather than to the administration, board of trustees, or other administrative office? Fordham Int Law J. 1993;5(2):115–7.
19. Otto IA, Breugem CC, Malda J, Bredenoord AL. Ethical considerations in the translation of regenerative biofabrication technologies into clinic and society. Biofabrication. 2016;8(4):042001.
20. Cook JA, McCulloch P, Blazeby JM, et al. IDEAL framework for surgical innovation 3: randomised controlled trials in the assessment stage and evaluations in the long term study stage. BMJ. 2013;346:f2820.
21. Ergina PL, Barkun JS, McCulloch P, Cook JA, Altman DG, Group I. IDEAL framework for surgical innovation 2: observational studies in the exploration and assessment stages. BMJ. 2013;346:f3011.
22. McCulloch P, Cook JA, Altman DG, Heneghan C, Diener MK, Group I. IDEAL framework for surgical innovation 1: the idea and development stages. BMJ. 2013;346:f3012.
23. Broekman ML, Carrière ME, Bredenoord AL. Surgical innovation: the ethical agenda: a systematic review. Medicine (Baltimore). 2016;95(25):e3790.

Evaluating Awake Craniotomies in Glioma Patients: Meeting the Challenge

Bart Lutters and Marike L. D. Broekman

Introduction

In 1886, the renowned English neurosurgeon Sir Victor Horsley (1857–1916) first used direct electrical stimulation to identify the primary motor cortex of a patient suffering from Jacksonian epilepsy [1]. The patient was awake while Horsley performed a craniotomy and subsequently applied faradic current to his brain. Following the initial report, awake craniotomies—combined with direct cortical stimulation—were soon adopted for the surgical management of glioma patients. Today, the procedure is regarded an effective method to identify indispensable brain regions, thereby minimizing the occurrence of postoperative functional deficits.

As evidence-based physicians, we owe to our current and future patients the continuous engagement in efforts to improve and evaluate the therapies we subject them to. The evaluation of awake craniotomies, however, confronts us with various scientific and ethical challenges. For one, the use of powerful research designs, such as the randomized controlled trial, may not be suitable due to a perceived lack of equipoise, lack of structured outcome collection, and small study populations.

B. Lutters
Department of Neurosurgery, Erasmus University Medical Center,
Rotterdam, The Netherlands

Department of Pediatric Neurosurgery, Brain Center Rudolf Magnus, University Medical Center Utrecht—Princess Máxima Center, Utrecht, The Netherlands

M. L. D. Broekman (✉)
Department of Neurosurgery, Computational Neurosciences Outcomes Center (CNOC), Brigham and Women's Hospital, Harvard Medical School, Boston, MA, USA

Department of Neurosurgery, Haaglanden Medical Center, The Hague, The Netherlands

Department of Neurosurgery, Leiden University Medical Center, Leiden, The Netherlands
e-mail: m.broekman@haaglandenmc.nl; m.l.d.broekman@lumc.nl

© Springer Nature Switzerland AG 2019
M. L. D. Broekman (ed.), *Ethics of Innovation in Neurosurgery*,
https://doi.org/10.1007/978-3-030-05502-8_12

Ethical concerns regarding the informed consent process, therapeutic misconception, and patient vulnerability further complicate awake craniotomy research.

In 2007, the Institute of Medicine introduced the learning health system (LHS), a system in which "knowledge generation is so embedded into the practice of medicine that it is the natural product of the healthcare delivery process and leads to the continuous improvement of care" [2]. Key components of the LHS include the search for alternatives to large randomized controlled trials, implementation of system databases, and fostering understanding of evidence-based medicine [2]. Here, we aim to identify the scientific and ethical challenges associated with awake craniotomy research in glioma patients and to propose the LHS and its associated ethics framework as a potential way to overcome some of these challenges.

Perceived Lack of Clinical Equipoise

The awake craniotomy procedure is commonly regarded as an effective method to minimize postoperative deficits and has, therefore, been widely adopted by the neurosurgical community. Despite its general acceptance, RCTs comparing the awake procedure to resection under general anesthesia have not been conducted. In contrast to the introduction of new pharmaceuticals, surgical innovation and research often take place outside controlled study conditions [3]; surgical techniques are commonly introduced with little or no ethical oversight, and if the opportunity for timely evaluation is not seized, the procedure may be adopted by the medical community without proper evidence [4]. This is also the case for awake craniotomies; since there is no genuine disagreement about the superiority awake craniotomies for the resection of gliomas within or nearby indispensable brain regions, most neurosurgeons feel that it would be unethical to expose patients to the procedure under general anesthesia instead [5–7]. Due to this perceived lack of clinical equipoise, large RCTs are not feasible, stressing the need for suitable alternatives.

The LHS has proposed several alternatives to the randomized controlled study design, including the cluster randomized trial (CRT) [2]. CRTs do not require randomization at patient level but allow participating institutions to perform their preferred standard of care. For instance, an institute which would favor glioma resection under general anesthesia would be compared to an institute preferring the awake procedure. This study design has several limitations, as the absence of randomization reduces the internal validity of the study. That being said, it seems preferable to supplement evidence from RCTs with high-quality nonrandomized studies, rather than solely relying on clinical opinion whenever randomized trials are not feasible [2].

Lack of Structured Data Collection and Sharing

Even though the awake craniotomy procedure—combined with direct cortical and subcortical stimulation—is widely regarded as the preferred treatment for gliomas within or nearby eloquent brain regions, technical details of the procedure vary significantly among institutions. Whereas some institutions reserve the procedure for tumors in proximity to essential brain regions (e.g., motor cortex, somatosensory cortex, language regions), others rely on awake craniotomies for most glioma resections, independent of tumor location [5]. Moreover, the extent of brain exposure, the use of intraoperative tasks, and the application of direct electrical stimulation vary among institutions and among surgeons from the same institution. These practice variations could—when properly registered and shared—be harnessed for the evaluation of the awake craniotomy procedure.

The LHS supports the implementation of large system databases and universal electronic health records, thereby providing a platform for continuous learning based on clinical decision-making. For resective glioma surgery, this would include the systematic registration of patient baseline information, tumor location, extent and type (awake versus general anesthesia) of craniotomy performed, direct stimulation protocol (including positive and negative stimulation sides), and treatment outcome in terms of postoperative functional impairment and tumor survival. Adequate registration and sharing of this data would potentially allow for the generation of evidence by means of observational treatment comparisons (OTC), in which outcomes from patients can be compared across institutions.

Informed Consent

Adequate informed consent for participation in any research activity should ideally consist of three components: (1) the surgeon must disclose evidence-based information regarding the risks and benefits associated with the procedure and reasonable alternatives, (2) the candidate subject should be capable of weighing the risks and benefits associated with the procedure, and (3) the candidate subject should voluntarily consent to the procedure, that is, the decision to participate should reflect his or her own opinion [8]. The informed consent process for awake craniotomy research is particularly challenging and hindered in various ways.

First, as no RCTs have been conducted on awake craniotomies, it is challenging for the neurosurgeon to disclose accurate information regarding the risks and benefits associated with the procedure. Disclosure becomes even more difficult when the surgeon wishes to investigate a modification to the existing awake craniotomy procedure, as the results of the modification cannot possibly be predicted. In such case, the neurosurgeon should always provide the information regarding his or her limited experience with the procedure. In addition, personal interests (e.g., career opportunity, financial gain, social status) may cause some involved in the evaluation of awake craniotomies to overstate the benefits and downplay the risks associated with the research activity.

The remaining two elements of informed consent—decisional capacity and voluntariness—give rise to additional ethical challenges, as glioma patients nominated for surgery represent a particularly vulnerable patient group. Indeed, it is conceivable that recently diagnosed glioma patients may not be in "the right psychological state of mind" to adequately weigh the risks and benefits associated with a certain research activity. The neurological nature of the disease and its therapy may further interfere with the patient's decisional capacity. Lastly, with regard to voluntariness, it has been shown that glioma patients often base their consent to a research activity on trust in their physician or on the idea that dissent would harm their doctor-patient relationship, rather than on the information disclosed by the neurosurgeon [8].

The LHS may help to overcome at least some of these challenges by "improving public understanding of the nature of evidence-based medicine [...] and the importance of supporting progress toward medical care that reflects the best evidence" [2]. "Patients participating in clinical research often misconceive a research activity to be a form of clinical care tailored to their individual medical needs." For instance, glioma patients may expect to receive certain benefits from participating in a cluster randomized trial or an observational treatment comparison, whereas, in reality, only future patients are likely to benefit [9]. Increased awareness of the nature of evidence-based medicine could potentially lessen this "therapeutic misconception" and smoothen the informed consent process, as informed patients would be able to take a general stance toward participating in research activities before the circumstances arise. In addition, the LHS encourages healthcare workers to similarly adopt an open attitude toward evidence generation and self-reflection, thereby potentially diminishing the influence of the personal interests on the informed consent procedure.

Discussion

Since the *Belmont Reports* of the 1970s, the field of bioethics has traditionally drawn a strict boundary between clinical research and care, the first being primarily concerned with the development of generalizable knowledge and the latter to benefit the individual patient [10]. Clearly, various scientific and ethical challenges are encountered in evaluating the awake craniotomy procedure in glioma patients, "stressing the need to rethink the way in which neurosurgical research and care is currently organized." We believe the LHS may provide a way to address the above-mentioned challenges while realizing that too much focus on learning activities may expose patients to disproportionate risks, abuse, and unjust distribution of burdens [11]. Consequently, Faden et al. have recently proposed an ethics framework to help ensure that research activities within a LHS are conducted in an ethically acceptable fashion [11].

The ethics framework by Faden and coworkers significantly departs from traditional bioethics in two ways: (1) the framework places a moral emphasis on learning and (2) sets a moral obligation to address unjust distribution of burdens within the healthcare system [11]. The moral obligation to learning includes both patients and healthcare professionals and holds that everyone involved in healthcare—both on the receiving and the providing end—had the moral responsibility to contribute to learning activities in order to enhance clinical practice "or the value, quality, or efficiency of the systems, institutions, and modalities through which health care services are provided" to the benefit of future patients [11]. This approach may somewhat temper traditional guidelines of ethical oversight and consent, thereby stimulating continuous learning activities to take place through the implementation of large system databases and data sharing.

The moral obligation to address unequal burdens within the healthcare system holds that healthcare workers and institutions have the responsibility to prevent that "the risks and burdens associated with a learning activity fall disproportionally on patients that are already disadvantaged" [11]. The widespread implementation of large system databases and universal electronic healthcare records carries with it the potential risk of taking unfair advantage of patients who are already particularly vulnerable to the undue influence of others. As mentioned earlier, glioma patients nominated for brain surgery represent a particularly vulnerable patient group, due to the severity and the disabling nature of their disease [12]. The obligation to address unjust inequalities will help to ensure that the burdens and benefits of continuous research activities will be equally distributed among patients and institutions "rather than placing the burden primarily on most desperate and refractory individuals" or on the less prosperous institutions.

Conclusion

Evaluating the awake craniotomy procedure in glioma patients confronts us with various scientific and ethical challenges. Here, we aimed to identify these challenges—namely, (1) a perceived lack of clinical equipoise, (2) a lack of structured data collection and sharing, and (3) an ineffective informed consent process—and to propose the concept of a learning health system as a potential solution. Firstly, the use of alternatives to large randomized studies—such as cluster randomized trials and observational treatment comparisons—proposed by the LHS may enhance awake craniotomy procedure research by filling in the gaps of evidence when RCTs are infeasible. The implementation of system databases and universal electric health records may further stimulate learning activities. Moreover, we believe that fostering an improved public and professional understanding of the nature of evidence-based medicine would smoothen the informed consent process and lessen the "therapeutic misconception." Finally, we believe that the ethics framework proposed by Faden and coworkers will minimize the risks associated with the blurring of the traditional boundaries between research and care while warranting the "do no harm" principle.

References

1. Greenblatt S. A history of neurosurgery: in its scientific and professional contexts. New York: Thieme; 1997. p. 23–5.
2. Olsen L, Aisner D, McGinnis JM. The learning healthcare system: workshop summary (IOM roundtable on evidence-based medicine). Washington DC: National Academies Press; 2007.
3. Broekman ML, Carrière ME, Bredenoord AL. Surgical innovation: the ethical agenda: a systematic review. Medicine (Baltimore). 2016;95(25):e3790.

 4. McCulloch P, Altman DG, Campbell WB, Flum DR, Glasziou P, Marshall JC, et al. No surgical innovation without evaluation: the IDEAL recommendations. The Lancet. 2009;374:1105–12.
 5. Brown T, Shah AH, Bregy A, Shah NH, Thambuswamy M, Barbarite, et al. Awake craniotomy for brain tumor resection: the rule rather than the exception? J Neurosurg Anesthesiol. 2013;25:240–7.
 6. Kirsch B, Bernstein M. Ethical challenges with awake craniotomy for tumor. Can J Neurol Sci. 2012;39:78–82.
 7. Serletis D, Bernstein M. Prospective study of awake craniotomy used routinely and nonselectively for supratentorial tumors. J Neurosurg. 2007;107:1–6.
 8. Bernstein M. Fully informed consent is impossible in surgical clinical trials. Can J Surg. 2005;48:271.
 9. Appelbaum PS, Roth LH, Lidz CW, Benson P, Winslade W. False hopes and best data: consent to research and the therapeutic misconception. Hast Cent Rep. 1987;17:20–4.
10. Kass NE, Faden RR, Goodman SN, Pronovost P, Tunis S, Beauchamp TL. The research-treatment distinction: a problematic approach for determining which activities should have ethical oversight. Hast Cent Rep. 2013;43:4–15.
11. Faden RR, Kass NE, Goodman SN, Pronovost P, Tunis S, Beauchamp TL. An ethics framework for a learning health care system: a departure from traditional research ethics and clinical ethics. Hast Cent Rep. 2013;43:16–27.
12. Ford PJ. Vulnerable brains: research ethics and neurosurgical patients. J Law Med Ethics. 2009;37:73–82.

Evaluation of Innovations in Neurosurgery

Ethical Considerations of Neuro-oncology Trial Design in the Era of Precision Medicine

Saksham Gupta, Timothy R. Smith, and Marike L. D. Broekman

Introduction

Precision medicine is a novel therapeutic paradigm that has revolutionized the medical management of many cancers. This paradigm is based on the premise that specific oncogenic mutations drive cancer growth and recurrence. Precision medicine therapies target the protein products of these mutations to modulate their effects. Many of these therapies have demonstrated improved outcomes for advanced disease, and many more are currently under study. The successes of precision medicine include vemurafenib, a B-Raf inhibitor, in *BRAF*-positive melanoma and gefitinib for *EGFR*-positive non-small cell lung cancer [1, 2]. These therapies have begun to be studied in neuro-oncological disease, and early case reports and small clinical trials have demonstrated its potential in the field [3–5]. Indeed, this novel approach has the potential to improve outcomes in devastating brain cancers where traditional chemo- and radiotherapy have had limited success. However, the alterations in clinical research methodologies bring ethical challenges. In this chapter, we review ongoing changes in trial design and pertinent ethical considerations.

Based on: Gupta S, Smith TR, Broekman ML. Ethical considerations of neuro-oncology trial design in the era of precision medicine. J Neurooncol. 2017 May 29.

S. Gupta · T. R. Smith
Department of Neurosurgery, Computational Neurosciences Outcomes Center (CNOC), Brigham and Women's Hospital, Harvard Medical School, Boston, MA, USA

M. L. D. Broekman (✉)
Department of Neurosurgery, Computational Neurosciences Outcomes Center (CNOC), Brigham and Women's Hospital, Harvard Medical School, Boston, MA, USA

Department of Neurosurgery, Haaglanden Medical Center, The Hague, The Netherlands

Department of Neurosurgery, Leiden University Medical Center, Leiden, The Netherlands
e-mail: m.broekman@haaglandenmc.nl; m.l.d.broekman@lumc.nl

Ethical Challenges of Early Phase Clinical Trials for Precision Medicine

Clinical trials of precision medicine therapy are currently being tested in patients whose cancers contain mutations that these therapies target. Next-generation sequencing [6] technologies have ushered a new era of individual tumor characterization by identifying specific oncogenic mutations within tumors. Cancer centers use different NGS technologies to varying degrees, and some centers, like the Dana-Farber Cancer Institute, screen all brain tumors for a predetermined of oncogenic mutations [7]. These methods aid in determining optimal adjuvant regimens for each tumor based on its molecular signature.

The specificity of precision medicine therapies makes them more personalized but also renders large randomized control trials (RCTs) difficult. Patients selected for these trials must share the same relevant genomic changes in addition to traditional inclusion criteria like age and the grade and stage of the cancer. The relatively low incidence of brain tumors also prohibits the recruitment of large cohorts. Furthermore, the high cost of these therapies may limit cohort sizes as well. The size, specificity, and expense of precision medicine will shift trial design away from RCTs and toward N of 1 and small cohort trials in neuro-oncology. This discussion focuses on the pertinent ethical challenges to this shift: informed consent, vulnerable patient selection, societal value, generalizability, institutional oversight, and justice.

Informed Consent

Informed consent is requisite when enrolling patients into all research protocols, including clinical trials. Neurosurgeons must explain the benefits and risks of a procedure or treatment to obtain sufficient informed consent. This communication must include the risks and benefits of precision medicine therapies and explain the basic premise behind precision medicine. However, physicians differ in their own understanding and experience with complex genomic underpinnings of these therapies, which may limit their ability to explain with adequate and accurate detail [8]. Patients also have various degrees of science and health literacy. These factors may combine to obfuscate patients' understanding of trials they may enroll into and may hamper extant communication barriers between patients and physicians [9, 10]. For precision medicine trials specifically, informed consent include consent to genetically sequence a tumor specimen and consent to medicate the patient in accordance with the research protocol.

Informed consent must explain and create options for the various outcomes of NGS screening. These include normal screens without any identified mutations, the identification of the mutations of interest for the trial, and incidental findings with known or unknown implications that may or may not be actionable. The complexity of genomics in neuro-oncology and mechanisms of targeted therapy may compound on the communication barriers that already exist between physicians and patients [6]. Furthermore, NGS exposes patients to discovering incidental genetic mutations of irrelevant and potentially unknown significance. The American College of Medical Genetics and Genomics' rescinded recommendation to broaden genomic screening variants in newborns without permission demonstrates the value patients hold in their right to not to know about genetic mutations [11]. This same ethical principle applies in oncology, in which patients have the right to not know about incidental mutations in their germline or tumor samples.

The four types of incidental findings in NGS are genetic variants that are medically actionable for the patient, medically actionable for the patient's family but not the patient, associated with disease but not currently medically actionable, and of unknown significance [12]. Patients vary in their desire to know each level of information. For instance, patients may desire to know about incidental findings that are medically actionable to prevent interventions that would improve their health, but not findings that are not actionable for them. Another patient may desire to know about findings that could benefit their family or findings that predispose them for other diseases to plan their life around this disease. These scenarios must be clearly delineated by the surgeon, and oversight committees should consider how to inform patients about the different types of incidental findings and how to create opt-in or opt-out systems for their disclosure.

The format of informed consent may be tailored to maximize patient understanding of the oncogenes of interest, targeted therapies, and potential incidental findings. Multidisciplinary teams can help design resources for patients and physicians to aid in this process; for example, genetic counselors may aid oncologists in communicating complex genetic information and implementing flexible consent processes [13]. Tiered informed consent involving multiple stages of obtaining consent provides patients time to understand the most central elements of the trial first while learning about less significant and more complex details like incidental findings at a later time [14]. The most central and necessary information about a trial is presented first, while less significant and more complex details like incidental findings are presented at an appropriate later time. Educational resources for patients may also be used to explain the uncertainty around incidental findings since patients often have difficulty understanding and responding to the presence of mutations with unknown clinical significance [15].

Vulnerable Patient Selection

Patients enrolled in experimental targeted therapy trials often have cancers that are refractory to standard therapeutic modalities. Their aggressive lesions render patients vulnerable to exploitation, so equitable participant selection is an important pillar in these trials. The primary goal of trials comprised of patients with refractory brain tumors should be to improve management of future patients with similar prognosis rather than patients with milder disease. For example, an *IDH* targeting therapy that demonstrates efficacy for end-stage glioblastoma patients must then enter therapeutic use for end-stage glioblastoma patients. Ensuring that the characteristics of trial enrollees reflect their intended beneficiaries becomes more difficult to accomplish in small cohorts, so care must be taken to carefully define the enrollee criteria and generalizability of a study. Furthermore, successful studies of therapies in a high-risk population will motivate the usage of the same therapy in lower-risk populations to benefit more patients, but patients with similar disease as the initial trial participants should continue to receive the therapy.

Societal Value

The societal value of precision therapies is a balance of initial cost and eventual patient outcomes. Targeted therapies require financial investment and time to develop but are ultimately applicable only to patients whose cancers harbor certain mutations [16]. There are cheaper alternatives to lower the burden of oncologic disease, including behavioral interventions such as smoking cessation, increasing exercise, and dietary change. However, this public health framework of oncology is not incompatible with a coexisting precision medicine framework. As behavioral initiatives combat risk factors like cigarette smoking and obesity, the average life span will continue to increase, and age is an unmodifiable risk factor for many cancers. These behavioral risk factors will likely have a larger impact on reducing the burden of cardiovascular disease than cancer, increasing the burden of cancer relative to cardiovascular diseases, which are currently the most common cause of death in the United States. Some cancers have no identified modifiable risk factors, and even for those that do, the avoidance of risk factors does not guarantee the prevention of cancer.

The societal value of targeted precision therapies may also extend beyond their immediate applications for certain cancers. One such example is "drug repositioning," the concept of applying therapies initially developed for one pathology to another. This concept holds greater potential in cancer, as similar oncogenes may drive multiple cancer types. For example, study of the V600E mutation in *BRAF* in melanoma led to the development of B-Raf inhibitors that years later have shown efficacy in case reports of craniopharyngioma [3]. The knowledge generated by NGS for early-phase targeted therapy trials will better define the genomic landscape of various brain cancers, thereby motivating future therapeutic study and contributing toward increased understanding of the biological significance of oncogenes more generally.

Generalizability

RCTs enroll dozens to hundreds of patients to increase statistical power for detailed statistical analysis on the effects of an experimental treatment. Smaller cohorts are conversely limited in the conclusions they may make with statistical confidence. A clinician determining whether to prescribe a targeted therapy to a patient based off the results of a two-enrollee trial may be unable to determine whether their patient will respond similarly to those enrollees. If those enrollees also responded differently to the therapy despite having the same mutational profile, the clinician cannot accurately predict which enrollee the patient would respond more similarly to.

Precision medicine aims to provide personalized therapies that maximize outcomes while minimizing off-target toxicity. However, physicians cannot determine potential toxicities without evidence from large and longitudinal cohort studies [17]. These unknown risks may be especially significant for intracranial pathologies given their interaction with such sensitive structures as the blood-brain barrier and eloquent cortex. A comprehensive understanding of the potential harms of any therapy requires longitudinal tracking beyond a clinical trial, so physicians may weigh its potential benefits against proxy approximations of risk, such as the toxicity associated with other therapies that have the same pharmacological target or by extrapolation of a therapy's toxicity profile in other cancers.

Oversight

Institutional oversight is a legal requirement for clinical research in the United States. Institutional Review Boards (IRBs) review all research protocols that use human test subjects, and no such protocol may be conducted without IRB approval. All phase II and III clinical trials must register with the National Institutes of Health to provide transparency to the public on these trials. IRB approval usually requires adequate justification for a trial, which may include supporting preclinical evidence. The specificity of targeted therapies renders the accumulation of sufficient preclinical evidence difficult, as in vivo testing would require animal models of cancers that specifically match the genomic signatures of a study population. The multiplicity of identified and yet-to-be identified oncogenes makes generation of all these animal models unfeasible. This problem will be compounded as precision therapies are developed to target multiple molecular aberrations in heterogeneous cancers. New experimentation protocols and emerging technologies such as ex vivo organoids and analytical data modeling can mitigate this problem, but ultimately, oversight committees must set strict and consistent evidential standards to warrant approval.

An increase in the quantity of early phase trials may supersede the capacity of IRBs to review and oversee all proposed trials; an oversight committee may follow a large trial with multiple co-sponsors more easily than many smaller, more specific trials more easily. Oversight committees may streamline operations to adapt to these anticipated changes, ensuring that sufficient oversight is present to maintain both the pace of innovation and the ethical soundness of research.

Justice

The concept of justice in clinical research requires that the conduct or results do not unfairly affect specific populations, especially marginalized groups such as patients of low socioeconomic status (SES). The selection criteria for early phase targeted therapy trials will include specific genetic data that presume access to genetic sequencing technologies and the expertise to interpret genetic data. These resources are most commonly located in central, advanced tertiary care institutions whose costs may be prohibitive to patients from low SES backgrounds [6]. After a trial, targeted therapies may be prohibitively expensive for low SES. Consequently, both the enrollees and beneficiaries of these trials are more likely to be patients of high SES, thereby unfairly excluding patients of low SES altogether.

Low representation of patients from low SES backgrounds may prevent the development of therapies specific for their cancers as well. These patients face different carcinogenic exposures, such as higher rates of heavy metal and cigarette smoke exposure, that may lead to different oncogenic profiles than patients of high SES. Potential therapies specific for genetic mutations that are more common in patients of low socioeconomic status due to certain environmental exposures may therefore never be developed, which would propagate these patients' exclusion from precision medicine. Patients from low SES backgrounds require representation in clinical trials to ensure their just inclusion in precision medicine [18, 19].

Conclusions and Future Directions

There has been a surge in early phase precision medicine trials to target mutations specific to individual cancers. These therapies hold immense potential to personalize therapeutic regiments to improve outcomes in neuro-oncology, but they come with unique changes to trial design, including their high costs and small cohort sizes. These changes require deliberate consideration of associated challenges in upholding clinical research ethics to protect patients and uphold scientific standards. The most salient ethical considerations include informed consent, vulnerable patient protection, societal value, generalizability, institutional oversight, and justice. While each of these present problems for early phase trials of targeted therapies, trial design can adapt to ensure ethical standards are met.

The increased difficulty of communicating complex information and handling incidental findings will make obtaining informed consent more difficult. Innovative informed consent structures and multidisciplinary approaches to educate patients will aid in this process [18]. In a recent cohort of cancer patients who underwent tumor sequencing with an NGS technology, 1% of reports revealed incidental findings [20]. This prevalence will vary according to the NGS methodology applied and cancer being studied; nevertheless, informed consent processes must include options for the patient to determine how to handle such scenarios. Tiered approaches and concrete example cases may help patients understand their options during the informed consent process.

The selection of trial enrollees will need to include protections to fairly treat vulnerable populations such as patients with end-stage diseases and those of low SES. Early trials for targeted therapies in cancer are often conducted in patients whose cancers have remained refractory to the standard of care, thus enrolling patients with more aggressive or even end-stage cancer. Protections should ensure these patients may drop out of a trial freely. Follow-up from these trials should ensure that the primary benefactors of the results of a trial reflect the trial's study population. These trials are expensive to conduct, and the therapies are expensive for patients to afford, but patients from low SES and other vulnerable populations should not be excluded.

The clinical and scientific benefits from targeted therapies should be recorded and analyzed to determine their true cost-effectiveness and societal impact. This can be done by unbiased third-party institutions with the expertise to conduct economic analyses in biomedical fields. Societal value also depends on researchers' clearly defining the generalizability of each trial as they become smaller and more specific, including specific mutations and demographic details. Trial organizers may collaborate with oncologists and epidemiologists to identify cancers without significant behavioral risk factors to focus study on these cancers.

Oversight committees will continue to review and monitor trial closely. The composition of oversight committees will benefit from representation by geneticists, oncologists, and biologists, all of whom contribute to the development of targeted therapies. Patients and their representatives may also provide input to determine the priorities and agenda of the development of precision medicine to increase public trust and involvement and improve equitability [21, 22].

While these are largely beyond the scope of this analysis, NGS and precision medicine raise challenging ethical questions regarding information sharing and management after the completion of trials. The integration of genomic results into patients' medical records must come with protections against genetic discrimination by employers, insurers, and other institutions [23]. Ethicists who analyze the social and professional underpinnings of this increasingly common form of discrimination need to coordinate with clinical investigators and policymakers closely to ensure such protections are in place. Furthermore, incidental findings with unknown clinical significance discovered by NGS in these trials may potentially be studied to determine their role in health and disease. The extent of potential prospective study of incidental findings should ideally be assessed prior to enrollment and included in informed consent protocols. Retrospective usage of such data should also require informed consent to protect patients from discrimination and respect their privacy.

We acknowledge the meritorious scientific background of early phase trials: decades of molecular biology and genomics research settings have yielded findings that have dramatically elongated patient survival and improved quality of life in cancer. As oncological management transitions to increasing specificity and personalization, ethical analysis associated with these changes is imperative to protect patients.

References

1. Chapman PB, Hauschild A, Robert C, Haanen JB, Ascierto P, Larkin J, et al. Improved survival with vemurafenib in melanoma with BRAF V600E mutation. N Engl J Med. 2011;364:2507–16.
2. Rosell R, Carcereny E, Gervais R, Vergnenegre A, Massuti B, Felip E, et al. Erlotinib versus standard chemotherapy as first-line treatment for European patients with advanced EGFR mutation-positive non-small-cell lung cancer (EURTAC): a multicentre, open-label, randomised phase 3 trial. Lancet Oncol. 2012;13:239–46.

3. Brastianos PK, Shankar GM, Gill CM, Taylor-Weiner A, Nayyar N, Panka DJ, et al. Dramatic response of BRAF V600E mutant papillary craniopharyngioma to targeted therapy. J Natl Cancer Inst. 2016;108:djv310.

4. Norden AD, Drappatz J, Wen PY. Targeted drug therapy for meningiomas. Neurosurg Focus. 2007;23:E12.

5. Prados MD, Chang SM, Butowski N, DeBoer R, Parvataneni R, Carliner H, et al. Phase II study of erlotinib plus temozolomide during and after radiation therapy in patients with newly diagnosed glioblastoma multiforme or gliosarcoma. J Clin Oncol. 2009;27:579–84.

6. McGowan ML, Settersten RA Jr, Juengst ET, Fishman JR. Integrating genomics into clinical oncology: ethical and social challenges from proponents of personalized medicine. Urol Oncol. 2014;32:187–92.

7. Wagle N, Berger MF, Davis MJ, Blumenstiel B, Defelice M, Pochanard P, et al. High-throughput detection of actionable genomic alterations in clinical tumor samples by targeted, massively parallel sequencing. Cancer Discov. 2012;2:82–93.

8. Murray MF. Educating physicians in the era of genomic medicine. Genome Med. 2014;6:45.

9. Ciardiello F, Adams R, Tabernero J, Seufferlein T, Taieb J, Moiseyenko V, et al. Awareness, understanding, and adoption of precision medicine to deliver personalized treatment for patients with cancer: a multinational survey comparison of physicians and patients. Oncologist. 2016;21:292–300.

10. Graham S, Brookey J. Do patients understand. Perm J. 2008;12:67–9.

11. Parens E. Drifting away from informed consent in the era of personalized medicine. Hast Cent Rep. 2015;45:16–20.

12. Bredenoord AL, Onland-Moret NC, Van Delden JJ. Feedback of individual genetic results to research participants: in favor of a qualified disclosure policy. Hum Mutat. 2011;32:861–7.

13. Everett JN, Gustafson SL, Raymond VM. Traditional roles in a non-traditional setting: genetic counseling in precision oncology. J Genet Couns. 2014;23:655–60.

14. Bradbury AR, Patrick-Miller L, Long J, Powers J, Stopfer J, Forman A, et al. Development of a tiered and binned genetic counseling model for informed consent in the era of multiplex testing for cancer susceptibility. Genet Med. 2015;17:485–92.

15. Fiore RN, Goodman KW. Precision medicine ethics: selected issues and developments in next-generation sequencing, clinical oncology, and ethics. Curr Opin Oncol. 2016;28:83–7.

16. Fleck LM. Just caring: assessing the ethical and economic costs of personalized medicine. Urol Oncol. 2014;32:202–6.

17. Lewis J, Lipworth W, Kerridge I. Ethics, evidence and economics in the pursuit of "personalized medicine". J Pers Med. 2014;4:137–46.

18. Budin-Ljosne I, Harris JR. Ask not what personalized medicine can do for you—ask what you can do for personalized medicine. Public Health Genomics. 2015;18:131–8.

19. Petersen KE, Prows CA, Martin LJ, Maglo KN. Personalized medicine, availability, and group disparity: an inquiry into how physicians perceive and rate the elements and barriers of personalized medicine. Public Health Genomics. 2014;17:209–20.

20. Bijlsma RM, Bredenoord AL, Gadellaa-Hooijdonk CG, Lolkema MP, Sleijfer S, Voest EE, et al. Unsolicited findings of next-generation sequencing for tumor analysis within a Dutch consortium: clinical daily practice reconsidered. Eur J Hum Genet. 2016;24:1496–500.

21. Bombard Y, Bach PB, Offit K. Translating genomics in cancer care. J Natl Compr Cancer Netw. 2013;11:1343–53.

22. Budin-Ljosne I, Harris JR. Patient and interest organizations' views on personalized medicine: a qualitative study. BMC Med Ethics. 2016;17:28.

23. Wauters A, Van Hoyweghen I. Global trends on fears and concerns of genetic discrimination: a systematic literature review. J Hum Genet. 2016;61:275–82.

The Ethics of Passive Data and Digital Phenotyping in Neurosurgery

14

Joeky T. Senders, Nicole Maher,
Alexander F. C. Hulsbergen, Nayan Lamba,
Annelien L. Bredenoord, and Marike L. D. Broekman

Passive Data in Healthcare and Neurosurgery

In the past few years, there has been a dramatic increase in the use of portable electronic devices including smartphones, tablets, smartwatches, and other wearables [1–15]. Traditionally, phones were used for calling and texting only, but the number of applications has grown exponentially, as well as the number of daily interactions we have with these devices. Portable electronic devices now have the digital sensors, processing speed, and memory to collect, share, and analyze a vast amount of data to provide granular insight into human behavior. For example, spatial trajectories, motions, cognitive capacity, sociability, and sleep cycles can all be derived from GPS, accelerometers, metadata on text and call activity, and screen on/off status, respectively [16–18]. These data streams are referred to as passive data (PD) since they are generated without any active participation of the subject as opposed to active data (surveys, audio samples, etc.). By collecting and analyzing PD, it is possible to make a very granular profile based on the behavior of the individual, the so-called digital phenotype. Digital phenotyping refers to the moment-by-moment

J. T. Senders · N. Maher · A. F. C. Hulsbergen · N. Lamba
Department of Neurosurgery, Computational Neurosciences Outcomes Center (CNOC),
Brigham and Women's Hospital, Harvard Medical School, Boston, MA, USA

A. L. Bredenoord
Department of Medical Humanities, Julius Center, University Medical Center Utrecht,
Utrecht, The Netherlands

M. L. D. Broekman (✉)
Department of Neurosurgery, Computational Neurosciences Outcomes Center (CNOC),
Brigham and Women's Hospital, Harvard Medical School, Boston, MA, USA

Department of Neurosurgery, Haaglanden Medical Center, The Hague, The Netherlands

Department of Neurosurgery, Leiden University Medical Center, Leiden, The Netherlands
e-mail: m.broekman@haaglandenmc.nl; m.l.d.broekman@lumc.nl

© Springer Nature Switzerland AG 2019
M. L. D. Broekman (ed.), *Ethics of Innovation in Neurosurgery*,
https://doi.org/10.1007/978-3-030-05502-8_14

"""

quantification of the in situ individual-level human phenotype, using data from smartphones and other personal devices.

Digital phenotyping may be particularly useful for numerous applications in the healthcare setting ranging from early symptom recognition to postoperative monitoring to the prediction of future trends. The PD used for digital phenotyping is a quantifiable, continuous source of longitudinal data and provides unique insight into the health and lifestyle of patients measured in their environment. Furthermore, PD is economically attractive, noninvasive, and scalable from individual patient care to clinical research, basic science, and the public health domain. In contrast, traditional ways of patient monitoring, such as surveys and routine hospital appointments, have always been subject to recall bias, subjective patient perception, variable physician interpretation, sporadic and episodic follow-up, and influence of the hospital environment in which outcomes are measured [19–21]. PD, by virtue of its objectivity, therefore has the potential to transform healthcare.

Although the collection and use of PD may be broadly applied across many domains of healthcare, neurosurgical patients in particular can benefit from digital phenotyping. Digital phenotyping denotes the quantification of behavior and functional outcome, and the human central nervous system is the biological machine underlying this. Pathological changes in the brain and spinal cord can result in functional and behavioral changes that can be captured by sensors in portable electronic devices. In fact, most digital phenotyping studies to date focus on psychiatric, neurological, or neurosurgical patients [4, 5, 13, 16, 17]. Additionally, neurosurgical interventions are temporally discrete, whereas neurologic and psychiatric interventions are often longitudinal, thereby allowing a distinct temporal assessment of the effect of surgery on functional outcome and behavior.

Commercial and academic institutions investigate numerous ways to improve health monitoring and effectively improve patient care; however, the sensitive nature of PD makes exploration of the field especially challenging. Given the rapid development of technologies that collect and analyze PD and the inevitable implementation in clinical practice, it is essential to have an overview of the ethical challenges that come along with it, as well as their potential solutions.

Ethics of Passive Data

Collection of PD

Informational privacy invasion is one of the leading concerns regarding the collection of PD [22–37]. Multiple factors can contribute to this loss of privacy. Due to its longitudinal and in situ nature, PD can be collected from a subjects' home and record intimate, private moments [22–26]. Data can even be collected on specific social or mobile activities that the subject does not intend to share [27, 29]. For example, GPS data collected for monitoring of physical functioning after surgery could also provide insight into specific locations patients visit and the time they spend there. Another factor that contributes to privacy invasion is so-called function creep. This means that the technology can be used beyond the purpose it was initially intended for. For example, the initial purpose of collecting PD might be to monitor clinical outcomes in operated patients, such as movement after orthopedic surgery. However, the same technology could then also be used by clinicians to check if the patient adheres to lifestyle intervention and whether they carry out enough physical exercise as part of a health plan. Through this, clinicians obtain a surveillance role. On the other hand, analyzing PD can also enhance patient privacy by allowing them to receive treatment in their own homes versus in a hospital setting [30].

Another frequently discussed concern regarding the collection of PD is informed consent [22, 24–27, 33, 38–43]. Many patients do not read or understand the consent forms. Instead, they automatically agree to terms and services [22, 25, 39, 41]. Due to the dynamic nature of PD collection and lack of insight into what knowledge can be derived from PD, patients may have less understanding and control over the data that is collected from them [24–27, 29, 33, 34, 36–38, 44–48]. Moreover, due to the rapid evolution of technology, the data that can be collected and the knowledge that can be derived from this data may prove to be different than at the timing of consent. Another issue identified is receiving informed consent from third parties or bystanders, such as family members or people living in the same household [26, 33, 38, 43]. Bystanders experience a loss of control over their data. Data can be unintentionally collected from them and sent to researchers without their knowledge or approval. Monitoring patients' texting behavior for longitudinal, cognitive assessment is an example because online conversations always include two or more people.

There are a variety of solutions that can improve informed consent procedures and also help minimize the extent to which the collection of PD infringes on a subject's privacy. Researchers could collect the simplest and minimum number of data elements necessary [26, 31–35]. Consent procedures could actively engage the individuals, thereby creating complete transparency in the data that is collected, the intent around the data acquisition, and the intentional and unintentional impact this may have on the patient [27, 33, 39, 40, 42]. This can be achieved through encouragement of an interactive dialogue, inclusion of visual aids, and stepwise verification of participant understanding, instead of a single episode explanation followed up by signing of the consent form [24, 33, 39, 40, 42]. Lastly, since the data is longitudinal and the technology is dynamic, it can be argued that the informed consent process should also be dynamic and flexible [22, 23, 26, 27, 29, 33, 36, 37, 43]. To help patients regain autonomy, it is suggested to give them as much control as possible over the technology and data [24, 25, 27, 33, 36–38, 45, 46, 49]. Privacy should be malleable to different contexts, and subjects should be able to change their personal privacy preferences at any time [22, 26, 29, 36, 37].

Use of PD to Improve Clinical Care

Care deliverers that collect and analyze PD are in a position to identify and act on health issues. Because healthcare interventions are highly reliant on the accuracy of this information, ethical concerns related to patient safety arise when PD directs clinical decision-making. For example, if a PD-based alarm system is responsible for detecting neurological deterioration after surgery, malfunctioning of this product can result in detrimental patient outcomes. Major themes include the regulation of devices, creating new safety systems, and ensuring safe PD-informed clinical decision-making [22, 24–27, 31, 33, 34, 36, 42, 43, 45, 47, 48, 50–52]. Currently, the evidence supporting the accuracy of apps processing PD is scarce, and there is no system in place to regulate their performance. Many healthcare apps are made directly available to the consumer through public app stores.

Digital phenotyping can reveal personal patient information that may lead to an increase in doctor biases. For example, a physician might notice that his or her patient is not adhering to prescribed lifestyle changes and as a result empathize less with the patient or underestimate the patient's complaints. Data that would have otherwise remained private can, therefore, result in stigmatization [27, 34, 47]. Additionally, an increase of monitoring by digital devices can come at the cost of real-life doctor-patient contact, causing depersonalization of healthcare. Much can be lost if human interaction is replaced with technology in the treatment of patients [43, 48]. For example, in the treatment of elderly patients, human touch and direct communication can have a significant impact on the well-being of the patient. Many elderly patients are already isolated, and replacing traditional healthcare interactions with technology will further exacerbate this issue [29]. Furthermore, the psychological effects of continuous surveillance by healthcare providers remain to be elucidated [28].

Incidental findings, findings that are discovered unintentionally and not related to the indication for which the data was originally investigated, are bound to occur with PD. For example, an accelerometer that tracks patients recovering from orthopedic surgery might also pick up a tremor related to an underlying Parkinson's disease. While incidental findings are inherent to medical tests, the added risk in PD lies in the fact that data is collected with such ease and in such large quantities that clinically relevant or irrelevant incidental findings are much more likely to surface at some point. Anticipating this is essential, both for overtreatment and avoidance of unwelcome situations of liability through negligence. Clarification is needed around whether and when to inform the patient and who is to inform the patient, especially regarding information that the patient may not have wanted to know in the first place [24, 25, 33, 52]. Lastly, disparities that exist can be made worse due to unequal access to technology for the collection and analysis of PD, resulting in an unequal benefit of PD-enhanced clinical care in the society [22, 24, 27, 29, 33, 36, 41, 43–45, 53, 54]. New technology is often unaffordable and inaccessible to lower-income populations [24, 53, 54]. Other vulnerable patient groups can be people with physical, disease-related, and mental impairments who might lack the technological skills required to use and benefit from personalized health technology [24, 29, 33, 44, 45].

A regulation system should be put in place to evaluate the safety and effectiveness of technology that can collect and analyze PD to enhance patient care [24, 26, 31, 33, 51]. All products should have to pass through a review board process before being approved for clinical use [24, 33, 51]. Some even suggest that mHealth products should meet FDA standards and health informatics standards to obtain clinical approval [51]. The Consolidated Standards of Reporting Trials (CONSORT) eHealth checklist might be useful for this purpose [40]. This checklist is an extension of the CONSORT statement specified for web-based or mobile health interventions or applications. The extent to which clinical decision-making can depend on patterns found in PD can vary enormously between different mHealth products. PD can, for example, be used as supplementary information for the clinician but can also be considered as an indicator for hospitalization or treatment. The required empirical evidence for the safety and effectiveness of PD products should, therefore, be balanced against the clinical impact of its application [40]. Not only should the significance and accuracy of the information be regulated but protocols for how to appropriately respond and take actions regarding patterns found in PD streams should also be implemented [50]. Incorporating continuous monitoring and evaluation of products after approval for clinical use is essential for establishing early detection of malfunctioning products [31]. Additionally, subjects should decide whether or not they would like to be informed about incidental findings [40]. Lastly, when products collecting and analyzing PD are implemented and standardized in clinical care, it is of paramount importance to ensure equal access, by making devices and software products affordable and settings understandable [22, 24, 27, 29, 33, 36, 41, 43–45, 53, 54].

Storage of Passive Data

Issues surrounding ownership and security are closely related to the storage of PD [26, 32, 33, 39, 42]. Currently, there are mixed perceptions about who owns PD. Patients, researchers, companies, hospitals, and academic institutions could all have a claim to the data. These claims are based on the type and degree of contribution involved in the research endeavor. For example, the patient is the subject of the data, researchers and companies may be the parties that analyze and derive insights from this data, thereby increasing the "value" of this data, and hospitals and academic institutions might be the ones that generate, store, and secure the data.

Security is of paramount importance for the storage of highly sensitive PD [22–25, 27, 31, 33, 36, 37, 39, 42, 43, 51, 55–57]. Many believe true security is not possible due to either government authorities or the failure of de-identification methods [22–25, 27, 31–33, 36–39, 42, 43, 51, 55, 57, 58]. Dependent on the (stability of a) legal framework in countries, authorities can demand access to PD [22, 24, 25, 32, 33, 38, 58]. If illegal activity is recorded during PD collection, courts in some countries might request access, and the security of the data will not be protected [22–25, 33]. Insufficient de-identification can also contribute to the loss of security. PD may be so rich in personal detail that the de-identification and encryption methods fail. Even if de-identification systems are put in place, combining multiple data points might still reveal an individuals' identity. PD results in a large volume of data, and current data storage systems may be outdated and insufficient to store all of this information safely. Lastly, PD is collected through mobile applications and wearable technologies allowing less computational power for advanced encryption methods than desktop computers [22–25, 27, 31, 33, 36, 37, 39, 42, 43, 51, 55–57].

Several solutions may help combat the issues of ownership and security for stored PD. Clarifying the details of ownership and making this known to participants is an important first step [25, 33]. Due to the variety of claims, complete ownership by one party might not be feasible or desirable; however, patients should have control over their data [24, 25, 27, 33, 36–38, 45, 46, 49]. The possibility of shared ownership between different parties might also be considered [44]. Researchers must protect data at the point of collection and have secure ways of transmitting the data. Also, regulations are needed regarding the security of PD storage systems [42, 43, 52, 54, 57, 59, 60]. These regulations should encompass quality standards for encryption methods but also specifications on who is allowed to access the data.

Secondary Use

PD has excellent potential for improving clinical care at the individual level, but on a larger scale, it can also provide valuable insight into disease development and progression in the scientific realm or public health domain. In the clinical realm, the trade-off between PD collection and improved patient care can be clear and transparent. In contrast, PD used for research or public health purposes is not directly beneficial for the individual patient who contributes their data.

In the scientific realm, it is suggested that the data collected be used solely for the purpose it was originally intended for [23, 25, 32, 58]. Upon conclusion of the study, the data must be deleted. If the data is to be used for a secondary purpose, informed consent must be received; however, this might be impossible in some cases since these patients might not be alive or within the scope of the hospital anymore [24, 33, 38]. It is harder to set boundaries for data access to governments and companies. Allowing for sharing of data between different institutions can promote the use of PD and increase the benefit we can derive from PD, but it is essential to clarify boundaries of access to data [23–25, 33, 58]. For example, data can be transmitted to technology companies to be analyzed by advanced quantitative methods; however, only the specific data elements that are crucial for performing the analysis should be shared.

The Big Picture

PD has the potential to make a tremendous impact on healthcare; however, this data is sensitive and personal, and many ethical challenges related to its collection, use, storage, and secondary use remain. Since this technology is rapidly and continuously improving, the increasing granularity of the collected data will push the ethical boundaries even further. Laying out a robust ethical framework will create an environment in which patients and their interests are protected, while still allowing for the benefits of PD to be harnessed for clinical, scientific, and public health goals.

Parallels with Ethical Concerns Outside the Realm of Passive Data

Despite the overall consensus on the ethical concerns surrounding PD, very few concerns are actually supported by empirical research. This lack of scientific evidence may be a practical consequence of the fact that PD has not yet been fully implemented in clinical practice. These ethical concerns are, however, closely related to those discussed in other domains. For example, the concepts of informed consent and patient privacy have already been explored since the early days of medical research, thereby providing valuable insight into the validity of the opinions, arguments, and suggested solutions reported in the ethical literature on PD.

One study reviewed the effectivity of different consent procedures and found that the traditional consent procedure, which consists of a brief explanation by a physician or researcher followed by signing of the consent form, is often incomplete [61]. Furthermore, this study showed that patients often had difficulties with recalling and demonstrating an understanding of the information, as well as with discussing this information with their caregivers. This review also found that simplified supplemental written materials, decision aids, educational videos, and utilization of the "repeat back method," in which patients repeat their understanding of what their caregiver has explained to them, were more effective communication tools [61]. Two other survey studies that investigated preferences related to informational privacy found that privacy concerns are neither static nor generalizable but differ between cultures [62] and are highly dynamic over time [60]. These cultural and temporal variations support the concerns regarding a single-time-point consent procedure and rigid, uniform privacy settings.

Ethical concerns related to PD also parallel those related to other data types.

Similar to PD, genetic data contains information with a high degree of granularity and specificity, which can cause issues with data anonymity and security. Additionally, insurance companies can refuse to cover patients based on patterns found in both passive and genetic data. To anticipate these issues, the American Society of Human Genetics makes a clear distinction between different levels of identification of data [63]. The level required for studies depends on the source of the material, the purpose of the study, and the extent to which additional information is necessary. Additional information, such as demographics, diagnosis, and family history, is not stored with the sample if it is not necessary to achieve the goal of the research [63].

The ethical concerns of PD are also similar those seen outside of the medical realm, such as those related to online behavior tracking by large companies such as Google, Facebook, and Verizon. Both PD and online behavior tracking have very low thresholds for generating and obtaining data, as they require neither invasive procedures nor active participation from subjects. Internet companies already collect data with tremendous ease, volume, and velocity, but this comes at the cost of transparency and informed consent. However, most people are unaware of the data that is collected from them [35, 38]. Lengthy and complex consent cause individuals to agree to terms, which they do not read or understand.

The vast and obscure collection of Internet data, however, has already triggered a response in the form of the right to data protection formulated in the General Data Protection Regulation that will be implemented and binding to all members of the European Union (EU) as of May 2018 [64]. This regulation aims to unify the regulatory environment for international businesses by advocating a "data protection by design." This unification strengthens the position of all citizens and residents of the EU, without heavily relying on the responsibility of the individual's informed consent. Although this regulation predominantly pertains to international businesses and companies presently, it has direct implications for the construction of an ethical framework on the use of PD in healthcare.

Constructing an Ethical Framework for PD

As of now, there is no clear ethical framework for PD, and as such, individual patients are not optimally protected. Change is both necessary and urgent: technology is evolving rapidly, and it should be expected that commercial enterprises will introduce PD applications shortly. Although PD is still a highly emerging field, the significant degree of consensus in the current ethical literature, the broader empirical evidence supporting the discussed concerns, and the parallels with ethical topics, solutions, and regulations in other medical and nonmedical fields can guide the construction of an ethical framework that ensures a safe and sound implementation of PD in healthcare.

We believe that portable devices that are used for PD collection provide opportunities for proper involvement of patients in the decision-making and data collection process. Smartphone applications can, for example, facilitate interactive and easy to understand consent procedures, as well as flexible control settings. One proposed solution to meet the concept of data protection by design, as well as flexible control settings, can be a default medium-level privacy setting that is adjustable to personal preferences. Establishing a transparent trade-off between the data that is collected and the individual or collective benefits that might be gained can function as a two-edged sword, both informing patients and encouraging them to support data-intensive medical research by contributing their data. Depending on the clinical impact, PD applications should be validated before implementation. Because their performance is dynamic and dependent on the training data, PD applications should be monitored and evaluated after implementation. In this way, deviating performance trends can be detected before they lead to worse patient outcomes.

Future Challenges

Although reviewing the ethical literature and drawing parallels with other fields can help us find solutions for many ethical challenges related to PD, solutions for other concerns remain to be elucidated. These unanswered questions are often induced by a conflict between two or more ethical, technical, or clinical values, as well as our inability to adequately predict how this field will develop itself.

While quality standards should be employed for the storage of PD, it is uncertain what these standards should look like or how they should be structured. This is partly due to the continuous arms race between data encryption and methods to break these encryptions. Luckily, these concerns mirror a global trend in which the importance of data security is increasingly acknowledged. A commonly proposed solution to safely store PD is data minimization: collecting the minimal number of data elements needed to achieve a specific purpose. However, this conflicts with commonly used big data approaches. The value or necessity of individual data elements is determined after they have been collected and analyzed; therefore, collected data points often have no specific purpose at the initiation of collection. Moreover, while security for data protection is a well-known issue, concerns regarding cybersecurity to ensure safety of PD products have been relatively underexposed. For example, malicious interference with the battery life or program settings of pacemakers could lead to detrimental outcomes [59]. Exploring the risks of these types of security breaches should have a high priority for future investigation.

Informed consent and data anonymity procedures are both subject to inherent limitations that do not guarantee effectiveness in all situations. The secondary use of medical research can make fully informed consent impossible, while data anonymity can hinder efficient reuse and linkage of data on a large scale, which is necessary for this research. "Broad consent" and "pseudonymization" are two frequently applied solutions to meet informed consent and anonymization criteria but represent an inadequate deployment of both concepts. Instead of stretching and thus devaluing consent and anonymization procedures, it may be better to shift away from a one-regulation-fits-all paradigm and acknowledge that different applications rely on different regulations. For example, some groups advocate that an exception can be made for informed concept in situations that serve the public interest (scientific research or the public health domain) where informed consent is virtually impossible (patients are already dead or out of the scope of the hospital when their data is used for secondary purposes). This concept is referred to as the research exemption. Acknowledgement that an exemption should be made in certain situations reduces the need to stretch ethical core values to fit all applications and effectively supports adequate deployment in situations where these concepts do apply [65].

Lastly, it remains to be elucidated whether the use of PD to optimize the delivered care also improves patients' quality of life. In the future, monitoring quality of life and patients' perception will reveal whether the advantages of PD outweigh the disadvantages, such as the psychological burden of continuously being monitored.

Conclusion

Although passively collected data has the potential to transform clinical management, several ethical challenges including privacy, informed consent, patient safety, ownership, storage, security, and secondary use of PD should be kept in mind.

A robust ethical framework will serve the dual purpose of protecting patients' interest and expanding the legitimacy of PD, with the ultimate goal of continuously improving clinical care.

References

1. Abdullah S, Matthews M, Frank E, Doherty G, Gay G, Choudhury T. Automatic detection of social rhythms in bipolar disorder. J Am Med Inform Assoc. 2016;23(3):538–43. https://doi.org/10.1093/jamia/ocv200.
2. Beiwinkel T, Kindermann S, Maier A, et al. Using smartphones to monitor bipolar disorder symptoms: a pilot study. JMIR Ment Health. 2016;3(1):e2. https://doi.org/10.2196/mental.4560.
3. Ben-Zeev D, Scherer EA, Wang R, Xie H, Campbell AT. Next-generation psychiatric assessment: using smartphone sensors to monitor behavior and mental health. Psychiatr Rehabil J. 2015;38(3):218–26. https://doi.org/10.1037/prj0000130.
4. Capecci M, Pepa L, Verdini F, Ceravolo MG. A smartphone-based architecture to detect and quantify freezing of gait in Parkinson's disease. Gait Posture. 2016;50:28–33. https://doi.org/10.1016/j.gaitpost.2016.08.018.
5. Ellis RJ, Ng YS, Zhu S, et al. A validated smartphone-based assessment of gait and gait variability in Parkinson's disease. PLoS One. 2015;10(10):e0141694. https://doi.org/10.1371/journal.pone.0141694.
6. Faurholt-Jepsen M, Frost M, Vinberg M, Christensen EM, Bardram JE, Kessing LV. Smartphone data as objective measures of bipolar disorder symptoms. Psychiatry Res. 2014;217(1-2):124–7. https://doi.org/10.1016/j.psychres.2014.03.009.
7. Faurholt-Jepsen M, Vinberg M, Frost M, Christensen EM, Bardram JE, Kessing LV. Smartphone data as an electronic biomarker of illness activity in bipolar disorder. Bipolar Disord. 2015;17(7):715–28. https://doi.org/10.1111/bdi.12332.
8. Grunerbl A, Muaremi A, Osmani V, et al. Smartphone-based recognition of states and state changes in bipolar disorder patients. IEEE J Biomed Health Inform. 2015;19(1):140–8. https://doi.org/10.1109/JBHI.2014.2343154.
9. Guidi A, Salvi S, Ottaviano M, et al. Smartphone application for the analysis of prosodic features in running speech with a focus on bipolar disorders: system performance evaluation and case study. Sensors (Basel). 2015;15(11):28070–87. https://doi.org/10.3390/s151128070.
10. Kostikis N, Hristu-Varsakelis D, Arnaoutoglou M, Kotsavasiloglou C. Smartphone-based evaluation of parkinsonian hand tremor: quantitative measurements vs clinical assessment scores. Conf Proc IEEE Eng Med Biol Soc. 2014;2014:906–9. https://doi.org/10.1109/EMBC.2014.6943738.
11. LeMoyne R, Tomycz N, Mastroianni T, McCandless C, Cozza M, Peduto D. Implementation of a smartphone wireless accelerometer platform for establishing deep brain stimulation treatment efficacy of essential tremor with machine learning. Conf Proc IEEE Eng Med Biol Soc. 2015;2015:6772–5. https://doi.org/10.1109/EMBC.2015.7319948.
12. Pan D, Dhall R, Lieberman A, Petitti DB. A mobile cloud-based Parkinson's disease assessment system for home-based monitoring. JMIR Mhealth Uhealth. 2015;3(1):e29. https://doi.org/10.2196/mhealth.3956.
13. Senova S, Querlioz D, Thiriez C, Jedynak P, Jarraya B, Palfi S. Using the accelerometers integrated in smartphones to evaluate essential tremor. Stereotact Funct Neurosurg. 2015;93(2):94–101. https://doi.org/10.1159/000369354.
14. Takac B, Catala A, Rodriguez Martin D, van der Aa N, Chen W, Rauterberg M. Position and orientation tracking in a ubiquitous monitoring system for Parkinson disease patients with freezing of gait symptom. JMIR Mhealth Uhealth. 2013;1(2):e14. https://doi.org/10.2196/mhealth.2539.

15. Wahle F, Kowatsch T, Fleisch E, Rufer M, Weidt S. Mobile sensing and support for people with depression: a pilot trial in the wild. JMIR Mhealth Uhealth. 2016;4(3):e111. https://doi.org/10.2196/mhealth.5960.
16. Onnela JP, Rauch SL. Harnessing smartphone-based digital phenotyping to enhance behavioral and mental health. Neuropsychopharmacology. 2016;41(7):1691–6. https://doi.org/10.1038/npp.2016.7.
17. Torous J, Kiang MV, Lorme J, Onnela JP. New tools for new research in psychiatry: a scalable and customizable platform to empower data driven smartphone research. JMIR Ment Health. 2016;3(2):e16. https://doi.org/10.2196/mental.5165.
18. Torous J, Onnela JP, Keshavan M. New dimensions and new tools to realize the potential of RDoC: digital phenotyping via smartphones and connected devices. Transl Psychiatry. 2017;7(3):e1053. https://doi.org/10.1038/tp.2017.25.
19. Schmier JK, Halpern MT. Patient recall and recall bias of health state and health status. Expert Rev Pharmacoecon Outcomes Res. 2004;4(2):159–63. https://doi.org/10.1586/14737167.4.2.159.
20. Aleem IS, Duncan J, Ahmed AM, et al. Do lumbar decompression and fusion patients recall their preoperative status? A cohort study of recall bias in patient-reported outcomes. Spine (Phila Pa 1976). 2017;42(2):128–34. https://doi.org/10.1097/BRS.0000000000001682.
21. McPhail S, Haines T. Response shift, recall bias and their effect on measuring change in health-related quality of life amongst older hospital patients. Health Qual Life Outcomes. 2010;8:65. https://doi.org/10.1186/1477-7525-8-65.
22. Balestra M, Shaer O, Okerlund J, Westendorf L, Ball M, Nov O. Social annotation valence: the impact on online informed consent beliefs and behavior. J Med Internet Res. 2016;18(7):e197. https://doi.org/10.2196/jmir.5662.
23. Markowetz A, Blaszkiewicz K, Montag C, Switala C, Schlaepfer TE. Psycho-informatics: big data shaping modern psychometrics. Med Hypotheses. 2014;82(4):405–11. https://doi.org/10.1016/j.mehy.2013.11.030.
24. Mikal J, Hurst S, Conway M. Ethical issues in using Twitter for population-level depression monitoring: a qualitative study. BMC Med Ethics. 2016;17:22. https://doi.org/10.1186/s12910-016-0105-5.
25. Miller G. The smartphone psychology manifesto. Perspect Psychol Sci. 2012;7(3):221–37. https://doi.org/10.1177/1745691612441215.
26. Mok TM, Cornish F, Tarr J. Too much information: visual research ethics in the age of wearable cameras. Integr Psychol Behav Sci. 2015;49(2):309–22. https://doi.org/10.1007/s12124-014-9289-8.
27. DiStefano MJ, Schmidt H. mHealth for tuberculosis treatment adherence: a framework to guide ethical planning, implementation, and evaluation. Glob Health Sci Pract. 2016;4(2):211–21. https://doi.org/10.9745/GHSP-D-16-00018.
28. Demiris G, Doorenbos AZ, Towle C. Ethical considerations regarding the use of technology for older adults. The case of telehealth. Res Gerontol Nurs. 2009;2(2):128–36. https://doi.org/10.3928/19404921-20090401-02.
29. Kelly P, Marshall SJ, Badland H, et al. An ethical framework for automated, wearable cameras in health behavior research. Am J Prev Med. 2013;44(3):314–9. https://doi.org/10.1016/j.amepre.2012.11.006.
30. Sorell T, Draper H. Telecare, surveillance, and the welfare state. Am J Bioeth. 2012;12(9):36–44. https://doi.org/10.1080/15265161.2012.699137.
31. Gimbert C, Lapointe FJ. Self-tracking the microbiome: where do we go from here? Microbiome. 2015;3:70. https://doi.org/10.1186/s40168-015-0138-x.
32. Landau R, Werner S. Ethical aspects of using GPS for tracking people with dementia: recommendations for practice. Int Psychogeriatr. 2012;24(3):358–66. https://doi.org/10.1017/S1041610211001888.
33. O'Doherty KC, Christofides E, Yen J, et al. If you build it, they will come: unintended future uses of organised health data collections. BMC Med Ethics. 2016;17(1):54. https://doi.org/10.1186/s12910-016-0137-x.

34. Prasad A, Sorber J, Stablein T, Anthony D, Kotz D. Understanding user privacy preferences for mHealth data sharing. In: mHealth multidisciplinary verticals; 2014.
35. Weinberg B, Milne G, Andonova Y, Hajjat F. Internet of things: convenience vs. privacy and secrecy. Bus Horiz. 2015;58(6):615–24.
36. Capon H, Hall W, Fry C, Carter A. Realising the technological promise of smartphones in addiction research and treatment: an ethical review. Int J Drug Policy. 2016;36:47–57. https://doi.org/10.1016/j.drugpo.2016.05.013.
37. Stip E, Rialle V. Environmental cognitive remediation in schizophrenia: ethical implications of "smart home" technology. Can J Psychiatr. 2005;50(5):281–91. https://doi.org/10.1177/070674370505000509.
38. Baldini G, Botterman M, Neisse R, Tallacchini M. Ethical design in the internet of things. Sci Eng Ethics. 2018;24(3):905–25. https://doi.org/10.1007/s11948-016-9754-5.
39. Kang HG, Mahoney DF, Hoenig H, et al. In situ monitoring of health in older adults: technologies and issues. J Am Geriatr Soc. 2010;58(8):1579–86. https://doi.org/10.1111/j.1532-5415.2010.02959.x.
40. Kaplan B. How should health data be used? Camb Q Healthc Ethics. 2016;25(2):312–29. https://doi.org/10.1017/S0963180115000614.
41. Paton C, Hansen M, Fernandez-Luque L, Lau AY. Self-tracking, social media and personal health records for patient empowered self-care. Contribution of the IMIA social media working group. Yearb Med Inform. 2012;7:16–24.
42. Savastano M, Hovsto A, Pharow P, Blobel B. Security, safety, and related technology—the triangle of eHealth service provision. Stud Health Technol Inform. 2008;136:709–14.
43. Smith RJ, Grande D, Merchant RM. Transforming scientific inquiry: tapping into digital data by building a culture of transparency and consent. Acad Med. 2016;91(4):469–72. https://doi.org/10.1097/ACM.0000000000001022.
44. Bietz MJ, Bloss CS, Calvert S, et al. Opportunities and challenges in the use of personal health data for health research. J Am Med Inform Assoc. 2016;23(e1):e42–8. https://doi.org/10.1093/jamia/ocv118.
45. Bramstedt KA. When microchip implants do more than drug delivery: blending, blurring, and bundling of protected health information and patient monitoring. Technol Health Care. 2005;13(3):193–8.
46. Cato KD, Bockting W, Larson E. Did i tell you that? Ethical issues related to using computational methods to discover non-disclosed patient characteristics. J Empir Res Hum Res Ethics. 2016;11(3):214–9. https://doi.org/10.1177/1556264616661611.
47. Essen A. The two facets of electronic care surveillance: an exploration of the views of older people who live with monitoring devices. Soc Sci Med. 2008;67(1):128–36. https://doi.org/10.1016/j.socscimed.2008.03.005.
48. Olff M. Mobile mental health: a challenging research agenda. Eur J Psychotraumatol. 2015;6:27882. https://doi.org/10.3402/ejpt.v6.27882.
49. Kelly D, Donnelly S, Caulfield B. Smartphone derived movement profiles to detect changes in health status in COPD patients—a preliminary investigation. Conf Proc IEEE Eng Med Biol Soc. 2015;2015:462–5. https://doi.org/10.1109/EMBC.2015.7318399.
50. Boye N. Co-production of health enabled by next generation personal health systems. Stud Health Technol Inform. 2012;177:52–8.
51. Draper H, Sorell T. Telecare, remote monitoring and care. Bioethics. 2013;27(7):365–72. https://doi.org/10.1111/j.1467-8519.2012.01961.x.
52. Kovach KA, Aubrecht JA, Dew MA, Myers B, Dabbs AD. Data safety and monitoring for research involving remote health monitoring. Telemed J E Health. 2011;17(7):574–9. https://doi.org/10.1089/tmj.2010.0219.
53. Klasnja P, Consolvo S, Choudhury T, Beckwith R, Hightower J. Exploring privacy concerns about personal sensing. In: Tokuda H, Beigl M, Friday A, Brush AJB, Tobe Y, editors. Pervasive computing. Berlin, Heidelberg: Springer; 2009. p. 176–83.

54. Kosta E, Pitkanen O, Niemela M, Kaasinen E. Mobile-centric ambient intelligence in health- and homecare-anticipating ethical and legal challenges. Sci Eng Ethics. 2010;16(2):303–23. https://doi.org/10.1007/s11948-009-9150-5.
55. Carter A, Liddle J, Hall W, Chenery H. Mobile phones in research and treatment: ethical guidelines and future directions. JMIR Mhealth Uhealth. 2015;3(4):e95. https://doi.org/10.2196/mhealth.4538.
56. Henderson EM, Law EF, Palermo TM, Eccleston C. Case study: ethical guidance for pediatric e-health research using examples from pain research with adolescents. J Pediatr Psychol. 2012;37(10):1116–26. https://doi.org/10.1093/jpepsy/jss085.
57. Pharow P, Blobel B. Mobile health requires mobile security: challenges, solutions, and standardization. Stud Health Technol Inform. 2008;136:697–702.
58. Lee LM, Heilig CM, White A. Ethical justification for conducting public health surveillance without patient consent. Am J Public Health. 2012;102(1):38–44. https://doi.org/10.2105/AJPH.2011.300297.
59. Kramer DB, Fu K. Cybersecurity concerns and medical devices: lessons from a pacemaker advisory. JAMA. 2017;318(21):2077–8. https://doi.org/10.1001/jama.2017.15692.
60. Prasad A, Sorber J, Stablein T, Anthony D, Kotz D. Understanding sharing preferences and behavior for mHealth devices. Institute for Security, Technology and Society, Dartmouth College; 2012.
61. Cordasco KM, Rockville MD. Making health care safer II. An updated critical analysis of the evidence for patient safety practices. In: Obtaining informed consent from patients: brief update review; 2013.
62. Morey T, Forbath TT, Schoop A. Customer data: designing for transparency and trust. Harv Bus Rev. 2015.
63. Godard B, Schmidtke J, Cassiman JJ, Ayme S. Data storage and DNA banking for biomedical research: informed consent, confidentiality, quality issues, ownership, return of benefits. A professional perspective. Eur J Hum Genet. 2003;11(Suppl 2):S88–122. https://doi.org/10.1038/sj.ejhg.5201114.
64. Rumbold JM, Pierscionek B. The effect of the general data protection regulation on medical research. J Med Internet Res. 2017;19(2):e47. https://doi.org/10.2196/jmir.7108.
65. Mostert M, Bredenoord AL, Biesaart MC, van Delden JJ. Big Data in medical research and EU data protection law: challenges to the consent or anonymise approach. Eur J Hum Genet. 2016;24(7):956–60. https://doi.org/10.1038/ejhg.2015.239.

Research Ethics: When Innovation Is Clearly Research

15

Nayan Lamba and Marike L. D. Broekman

Introduction

While the distinction between research and patient care is not always clear, it is the responsibility of clinicians and scientists to rigorously evaluate and apply a structured framework within which to appraise any research that is to be conducted on patients. The importance of this responsibility cannot be overemphasized. History has demonstrated to us the atrocities and violations of human dignity that can occur in the setting of unethical research.

Such critical appraisal has certainly become more challenging in the last few decades with the rapid advancements in technology that have tremendously expanded what can and cannot be done. This is where applying structured ethical frameworks serves the research community well. Emanuel et al. described seven requirements for ethical research that elegantly capture the essence of what makes research ethical [1]. These include social or scientific value, scientific validity, fair subject selection, favorable risk-benefit ratio, independent review, informed consent, and respect for potential and enrolled subjects.

In this chapter, we aim to explore these seven ethical principles in a neurosurgical case, utilizing human head transplantation as an illustrative example through which to explore the concept of when innovation in neurosurgery is purely experimental.

N. Lamba
Department of Neurosurgery, Computational Neurosciences Outcomes Center (CNOC), Brigham and Women's Hospital, Harvard Medical School, Boston, MA, USA

M. L. D. Broekman (✉)
Department of Neurosurgery, Computational Neurosciences Outcomes Center (CNOC), Brigham and Women's Hospital, Harvard Medical School, Boston, MA, USA

Department of Neurosurgery, Haaglanden Medical Center, The Hague, The Netherlands

Department of Neurosurgery, Leiden University Medical Center, Leiden, The Netherlands
e-mail: m.broekman@haaglandenmc.nl; m.l.d.broekman@lumc.nl

© Springer Nature Switzerland AG 2019
M. L. D. Broekman (ed.), *Ethics of Innovation in Neurosurgery*,
https://doi.org/10.1007/978-3-030-05502-8_15

Case: Head Transplant

Since the turn of the last century, the prospect of head transplantation has captured the imagination of neurosurgeons, scientists, and the general public. Recently, head transplant has regained attention in popular media, as neurosurgeons have proposed performing this procedure in 2017. An international team based in Italy has proposed performing the first human head transplant on a patient suffering from a muscle wasting disease that has essentially destroyed his body and left him wheelchair-bound [2]. Their protocol, based on HEAVEN, the "head anastomosis venture" project, indicates that they will decapitate the heads of two men, a donor and a recipient, and subsequently transplant the recipient's head onto the donor's body via an elaborate vessel anastomosis and spinal fusion protocol [2]. The protocol involves a sharp, highly controlled severance of the spinal cords followed by the use of "fusogens," a waxy substance made of PEG, to fuse the donor and recipient spinal cord stumps together. This all must be done in less than 60 min to avoid risk of irreversible ischemia to the recipient's brain [2].

The patient who is set to undergo the procedure reached out to the team himself when he learned of their plans to perform this surgery in humans [3, 4]. He is a 31-year-old computer scientist who suffers from Werdnig-Hoffman disease [4]. His disease has been progressing since birth, and he now has minimal, if any, control over his body. He sees this operation as his last attempt at saving his life [4].

Seven Principles of Ethical Research

Social or Scientific Value

In order for a research trial to meet the requirement of social or scientific value, it must have the potential to lead to improvements in health and well-being or test a hypothesis that could increase knowledge about human biological systems [1]. The proposed human head transplantation experiment easily satisfies the latter condition insofar as it will offer scientists the opportunity to learn about recovery of function in patients with spinal cord injury and subsequent fusion. Spinal cord fusion following transection remains one of the technical challenges of human head transplantation [5]. To date, numerous animal experiments have utilized PEG, or polyethylene glycol, as a fusogen to help accelerate the process of fusion and functional restoration in a severed spinal cord [6]. Such rodent experiments have demonstrated rodents' ability to recover feeding, ambulation, and other normal motor behaviors within days to weeks following treatment with PEG. However, while this data demonstrates that a severed spinal cord can regain function, these experiments must be replicated in human studies before any definitive conclusions about their applicability in the clinical realm can be established [6]. Therefore, the proposed human head transplantation, whether it fails or succeeds, would serve to increase the knowledge about spinal cord recovery in humans, a field where knowledge is currently largely theoretical.

However, whether the protocol meets the first requirement of improving human health and well-being is less clear. We think that it is unlikely that the patient who undergoes the head transplant procedure will even survive the procedure for long [7]. Indeed, upon evaluating the success of head-body transplants thus far, the prospects appear bleak. In 2015, in mice head-body transplants performed by scientist Dr. Xiaoping Ren, one of the collaborators of the Italian neurosurgeon leading the human head transplant project, only half survived greater than 24 h, with the longest survival being 6 months [8]. Moreover, according to Ren, experiments in rats have only had a 30–50% survival rate. In addition, a head transplant performed on a monkey in China did not allow for recovery of movement after the procedure [9]. Thus, even if the patient does survive following the procedure, how long he will survive and whether he will recover motor function is unclear, making fulfillment of the first ethical criteria weak at best.

Scientific Validity

The second principle that must be fulfilled for a human research trial to be considered ethical is that it must be conducted in a methodologically rigorous manner with proper scientific design and that it must be practically feasible [1]. As stated above and outlined in a recent review published on human head transplantation [5], there are at least four technical challenges that remain to be fully addressed in human head transplantation, including spinal cord transection and anastomosis; spinal cord fusion; revascularization, neuroprotection, and cerebral ischemia; and pain control [5]. Experiments on spinal cord transection and fusion thus far are based upon animal models that do not replicate human physiology; moreover, injuries to which fusogens have thus far been applied are not equivalent to those that would occur in human head transplantation [5]. In terms of prevention of postoperative pain, preliminary studies have explored the use of a selective lesion in the subparietal white matter to target the sensory component of chronic pain that might develop following the transplant [10]. However, members of the Italian team leading the human head transplant project have themselves stated that further research is warranted before such methods can be applied for chronic post-transplant central neuropathic pain.

Therefore, the head transplantation protocol as it currently stands is not based upon a sound, scientifically valid design, having been tested in animal models that do not replicate human physiology. From a practical standpoint, it is also not yet completely feasible, in that there are technical challenges that remain, such as the potential for cerebral ischemia, which have not been fully addressed.

One of the challenges of performing research in humans is that scientists cannot test preliminary protocols on humans, as they might be able to do in experiments with other animals. So, this begs the question of how a human research trial is ever truly ready to be performed on a human. The aforementioned Italian team acknowledges that further studies in human cadavers need to be performed before a human head transplant can be undertaken on an actual live human being [9]. As of now, however, such extensive experimentation on human cadavers has not been performed, undermining the scientific validity of the current protocol and its applicability and potential for success in humans.

Fair Subject Selection

The third criterion put forth by Emanuel is that of fair subject selection. Fair subject selection requires that the goals of the study be the primary basis for determining who will be recruited and enrolled in the study. Vulnerability, privilege, and other such factors should not influence the subject selection. In addition, the results of the study need to be generalizable to the populations that will use the intervention. Along similar lines, the individuals who bear the risks and burdens of the research should also be in a position to enjoy the procedure's benefits [1].

According to the team working on this project, the ideal patient for a head transplantation procedure is someone young suffering from a condition that devastates the body but leaves the mind and brain intact, such as a muscular dystrophy [11]. The patient who is scheduled to undergo the first human head transplant in 2017 is a 31-year-old man who suffers from Werdnig-Hoffmann disease, a type of genetic spinal muscular atrophy that has rendered him wheelchair-bound his entire life. He is only able to feed himself, steer his wheelchair, and type. Moreover, his condition is fatal, and doctors are unsure how much time he has left to live [12]. Performing the surgery on a patient like this would certainly be in line with the goals of the study, since the head transplantation is meant for patients without a functioning body. He would fulfill the criteria of being someone who would stand to benefit from the intervention should it succeed. Moreover, should the surgery succeed, it would have implications for treatment of future patients with muscular atrophy. However, whether or not Spiridonov is vulnerable is a much more difficult question to answer.

Numerous studies have found links between spinal cord injury and depression, with rates of depression in these patients significantly higher than in the general medical population, ranging between 18.7 and 26.3% [13]. Dr. David Stevens, CEO of the Christian Medical and Dental Associations, questioned whether patients suffering from traumatic brain or spinal cord injuries can truly give informed consent for a surgery like head transplantation. He brings attention to the links between depression and suicidal ideation in patients who are quadriplegic, as well as acknowledges the desperation that such patients might experience in search of some sort of cure [12].

Therefore, unless we develop a method for rigorously screening patients who wish to undergo a head transplant for depression and ensure that their current medical situation has not made them vulnerable to make such a decision, human head transplantation does not clearly fulfill the criteria of fair subject selection. In the field of neurosurgery, conditions inherently affect the brain and can have a devastating impact of an individual's quality of life. Neurological illnesses therefore are more likely to affect an individual's decision-making capacity and mental wellness than pathologies in any other medical field. As neurosurgeons, it is therefore our duty to take extra caution to ensure that fair subject selection occurs in our clinical research.

Favorable Risk-Benefit Ratio

The fourth principle that a human research trial must fulfill to be considered ethical is that it must have a favorable risk-benefit ratio. For this element to be fulfilled, three criteria must be met: potential risks to individual should be minimized, potential benefits to individual subjects should be enhanced, and potential benefits to individual subjects and society should be proportionate to or outweigh the risks [1]. Human head transplantation does not fulfill the first of these; minimization of risks is done by using procedures with sound research design and/or by using procedures already performed on the subjects for diagnostic or treatment purposes. However, as described in the section on scientific validity, many of the steps involved in the human head transplantation protocol are not based on a validated research design and are currently highly experimental.

Human head transplantation also does not clearly meet the second of these criteria, i.e., that of delineating and enhancing benefits to individual subjects. This is because the protocol of head transplantation is so theoretical to begin with that whether it will even offer *any* benefit to patients undergoing the procedure is currently unclear.

Finally, the third piece of this principle is that the potential for benefits should be proportionate to or outweigh the risks. While this is a challenging standard to quantify, Emanuel identifies it as one of the utmost importance, as it "embodies the principles of nonmaleficence and beneficence, fundamental values of clinical research" [1]. Based upon the discussions above, the risks of the procedure (potential death of the patient) clearly outweigh the benefit of motor recovery, which is purely hypothetical at this stage.

Therefore, human head transplantation also does not fulfill the criteria of a favorable risk-benefit ratio.

Independent Review

The fifth element of an ethical human research trial is that of independent review. Independent review is essential to minimize the impact of conflicts of interest that inevitably arise when investigators conduct research that may help their patients, may allow them to obtain funding, and/or may possibly advance their careers [1]. Independent review also acts as a check to ensure that the research being proposed meets the ethical elements discussed in this chapter and that its potential for benefits outweighs the potential risks [1]. Moreover, independent review has a role in social accountability, assuring that patients are not being exposed to harms for the benefit of society [1].

Currently, it is unclear whether the team planning to perform the head transplant has submitted or plans to submit their protocol for review by an independent review board [7]. The procedure is currently slated to be performed where the team is based in China, but the Chinese government has not yet approved the surgery [12]. As pointed out in the European Association of Neurosurgical Societies (EANS) ethical committee's statement, the Chinese government has a history of less strict ethical oversight and regulation [7]. China allows scientists more leeway in medical research than Western countries [3]. For example, it has fewer restrictions on cloning and has harvested organs from executed prisoners in the past [3]. Recently, China caused some controversy for genetically editing human embryos with *Crispr*, as well [3]. Therefore, even if the Chinese government does approve the surgery, the decision may still not be ethically sound and warrants further independent review.

A surgery with risks as serious as the ones that this one carries and one that is almost entirely based upon extrapolations from mostly unsuccessful animal experiments warrants the utmost scrutiny. To date, the proposed surgery does not fulfill the criteria of independent review.

Informed Consent

Arguably the most important of the ethical requirements of clinical research is informed consent. Emanuel describes informed consent as having two purposes: first, ensuring that individuals control whether or not they enroll in a clinical trial and, second, participating only when the research is in line with their personal values and interests [1]. In the case of the head transplant surgery, the patient on whom the surgery is planned for is said to have reached out to the surgeon interested in the procedure himself [3]. The patient initially became interested in the work of Dr. Robert White, a surgeon of the 1970s, who transplanted the heads of rhesus monkeys onto others' bodies; when he heard that a surgeon in Italy had plans to perform such a surgery in humans, he is the one who sent the surgeon an email volunteering for the procedure [3]. He made the decision to participate himself and has expressed the belief that this procedure is in line with the interests he has for himself: "Removing all the sick parts but the head would do a great job in my case...I couldn't see any other way to treat myself" [3].

However, just because an individual volunteers for a procedure does not necessarily mean that his consent is informed. In addition to making a voluntary and uncoerced decision to participate, individuals must also be accurately informed of the purpose, risks, benefits, and alternatives of the research; they must also demonstrate an understanding of this information and how it would affect their own clinical situation [1]. In the case of head transplant, patients need to be properly informed about the theoretical nature of the surgery and that, as detailed above, it has not been previously performed in humans; animal models to date have demonstrated poor survival; and physicians cannot accurately predict how he will respond to the transplanted, foreign body. It is unclear whether the team has formally discussed all of the uncertainties and possible outcomes following the surgery with Spiridonov. In order for head transplant to meet the criteria of informed consent, it is imperative that these details are discussed in great detail with any patient who volunteers for the procedure.

Moreover, each patient's ability to provide informed consent, based on his or her mental capacity, should also be formally assessed. Only after each patient is formally evaluated for his or her decision-making capacity, the risks and benefits of the surgery are outlined to that patient, and the applicability of the surgery to the individual is explained can head transplant fulfill the requirement of informed consent.

Respect for Potential and Enrolled Subjects

Finally, the seventh ethical requirement that must be fulfilled by the head transplant surgery is that of respect for subjects. This encompasses the following five features: (1) privacy of subjects must be maintained; (2) subjects must be allowed to withdraw without penalty; (3) should new information about the patient's clinical condition or about the intervention, including new risks or benefits, arise during the study, the patient should be informed about this new information; (4) patients' conditions should be monitored throughout the study, and they should be appropriately treated or removed from the study if adverse reactions occur; and (5) patients should be informed about any information that was learned from the research being performed [1].

Because of the unprecedented, radical nature of the head transplant surgery, the procedure has gained international media attention. The involved patient's confidentiality has therefore not been maintained, and he has even stated, "I'm really, really tired of being famous… It's exhausting, and it takes a lot of your time, for nothing" [3]. If head transplantation were to be performed on other patients in the future, more care should be taken to protect the privacy of the patients who are undergoing the surgery. However, from a practical standpoint, this will be highly challenging unless head transplantation becomes a standard medical procedure.

The patient is allowed to change his mind to undergo the procedure at any time [12], but practically, once the procedure actually begins, this will not be feasible. As per the current plan, he would be kept comatose with barbiturates and other drugs for weeks after the surgery. Machines would be utilized to assist with his breathing and blood flow during this period. Upon signs of motor recovery, the doctors would taper the drug regimen and allow him to regain consciousness [3]. Therefore, once the surgery begins, Spiridonov would have no way to back out.

Along the same lines, if new information were learned during the surgery, conveying this information to Spiridonov would likely be of little benefit if his spinal cord has been transected and he has been drugged. Should adverse reactions, such as rejection of the donor body, occur during the procedure, again, it is unclear if there is anything that the surgeons would be able to do at that point to salvage or reverse any damage that had occurred to the patient.

The nature of this surgery is such that once it begins, there is little that can be done to involve the patient in any decision-making that needs to occur throughout the procedure. Moreover, it is unclear whether the surgeons have a detailed monitoring or backup plan in place should things go awry in the middle of the procedure. Based on these realities, head transplantation does not demonstrate respect for its subjects.

Case

As it currently stands, head transplantation does not meet the ethical requirements to be performed on a human. While this surgery has the potential to provide new information about how humans recover after spinal cord injury and therefore guide the development of future treatments for such patients, it is simply too experimental in its current state to fulfill basic ethical tenets, such as those of scientific validity, a favorable risk-benefit ratio, and respect for persons. Moreover, if plans to move forward with such an experimental surgery do move forward, more strict regulations to ensure proper informed consent and thorough oversight by an ethical institutional review board are necessary.

Conclusion

Neurosurgeons take care of extremely sick patients who suffer from devastating illnesses that can severely impact their daily functioning and quality of life. While it is tempting to get drawn by the potential that technical innovation offers, as a community, neurosurgeons must be more cognizant than ever before to ensure that we are maintaining the highest standards for any new procedure or surgery that we perform on a patient. Human dignity, patient beneficence, and nonmaleficence are fundamental values that should be at the forefront of every clinical innovation, and no innovation, no matter how promising, should be performed on a patient as purely experimental. When innovation is purely research, it is not yet ready to be tested in the clinical realm if it does not meet the seven requirements for ethical research.

References

1. Emanuel EJ, Wendler D, Grady C. What makes clinical research ethical? JAMA. 2000;283:2701–11.
2. Kirkey S. Meet Sergio Canavero, the brain behind the first head transplant and the HEAVEN project. National Post. 2016.
3. Kean S. The audacious plan to save this man's life by transplanting his head. The Atlantic. 2016.
4. Will Stewart NF. EXCLUSIVE: revealed, the terminally ill man set to be first to undergo the world's first full HEAD transplant pioneered by doctor branded 'nuts'. DailyMailcom. 2015.
5. Lamba N, Holsgrove D, Broekman ML. The history of head transplantation: a review. Acta Neurochir. 2016;158(12):2239–47.
6. Canavero S, Ren X. Houston, GEMINI has landed: spinal cord fusion achieved. Surg Neurol Int. 2016;7(Suppl 24):S626–8.
7. Brennum J, Ethico-legal Committee of the European Association of Neurosurgical Societies. The EANS Ethico-legal Committee finds the proposed head transplant project unethical. Acta Neurochir. 2016;158(12):2251–2.
8. Ren XP, et al. Head transplantation in mouse model. CNS Neurosci Ther. 2015;21(8):615–8.
9. Gray R. Doctor planning world's first head transplant says he is preparing for his 'Frankenstein' surgery by REANIMATING human corpses. 2016.
10. Canavero S, Bonicalzi V. Central pain following cord severance for cephalosomatic anastomosis. CNS Neurosci Ther. 2016;22(4):271–4.
11. Canavero S. HEAVEN: the head anastomosis venture project outline for the first human head transplantation with spinal linkage (GEMINI). Surg Neurol Int. 2013;4(Suppl 1):S335–42.
12. Blair L. Man volunteers for first human head transplant neurosurgeon calls 'HEAVEN'. 2016.
13. Williams R, Murray A. Prevalence of depression after spinal cord injury: a meta-analysis. Arch Phys Med Rehabil. 2015;96(1):133–40.

Part IV

Innovation in Neurosurgery: Required Culture and Team Collaboration

Innovation and Team Collaboration in Neurosurgery

16

Saskia M. Peerdeman

Introduction

Healthcare is becoming an increasingly complex endeavor, requiring expertise and adequate actions from all healthcare professionals including a variety of medical specialists. In order to achieve a high standard of patient care, both interdisciplinary and multidisciplinary teams are formed every day in routine and emergency settings. Since the publication of the National Institutes of Health (NIH) report "To err is human" in 2000, there has been a growing awareness that a substantial number of human errors, accidents, and complications occur in patient care [1]. In the USA, additional medical costs due to such errors are estimated at a staggering $17 billion per year [2]. These errors and accidents occur even when the level of expertise of individual team members is exceptionally high. Indeed, this could indicate that the overall performance of the team may be significantly compromised by ineffective teamwork, poor communication, and other nontechnical skills such as situational awareness, decision-making, leadership, and assertiveness. This is also true in surgery. An analysis of errors in a selection of hospitals in the Netherlands revealed that 80% of errors in a surgical setting could be attributed to human factors [3]. Setting and maintaining a high standard of team performance are therefore critical in a progressively complex healthcare system. The Royal College of Physicians and Surgeons of Canada has summarized key competencies (the so-called CanMEDS Roles) to include medical expert (as the central role), communicator, collaborator, health advocate, leader, scholar, and professional [4]. Placing careful emphasis on these roles, raising awareness of their potential impact, and providing adequate training along these axes may improve (surgical) team performance and therefore patient care.

S. M. Peerdeman (✉)
Department of Neurosurgery, Amsterdam University Medical Centers,
Amsterdam, The Netherlands
e-mail: sm.peerdeman@vumc.nl

© Springer Nature Switzerland AG 2019
M. L. D. Broekman (ed.), *Ethics of Innovation in Neurosurgery*,
https://doi.org/10.1007/978-3-030-05502-8_16

153

- *Communication*. Examples include negotiating, working collaboratively with the information given, openness of communication, quality of communication, and specific communication practices; it is important to note that deficiencies in the quality of the communication can endanger the patient's safety. Being a skilled specialist does not necessarily equate to being a skilled communicator.
- *Leadership*. Examples include leadership style and adaptive leadership behavior. With good leadership the roles and responsibilities within a team become clear, and the expertise of all members can be optimally exploited.

Innovation and Teams

In the last decennia, many technical innovations in neurosurgery have been introduced. For example, neuronavigation, endoscopic techniques, and various new spinal stabilization techniques are now routinely used, with many other technical innovations on the way. The introduction of such new technologies is, in general, a stepwise process to move a technical innovation from concept to development and implementation to routine clinical practice.

In the hospital, surgeons need to be allowed to apply their newly acquired knowledge and ideas about how to innovate. Team members will need the necessary space to make a suggestion and to debate this in a constructive manner; this will require an open culture and a safe working environment. Enhancing knowledge cannot occur without reflection. An unsafe environment, such as one where downsizing the organization is a looming threat, will slow if not crush innovation.

After the introduction of an innovation, surgeons will develop experience using the innovative treatment or device. Sharing of the individual experience will increase the collective knowledge about an innovation. A positive effective presence of a team leader (such as the head of department) plays a key role in information sharing within the team. This in turn will stimulate further team innovation [13].

Traditionally, a neurosurgical patient is treated by a monodisciplinary team, where its members tend to have similar functional knowledge and conduct similar clinical tasks. Given the more equal status among team members, such monodisciplinary teams tend to have lower authority differentiation, complicating overall team performance.

However, with ongoing developments in radiosurgery and endovascular treatments, there is a continued need for collaboration across disciplines. Introducing an innovation from a different discipline into neurosurgical practice requires an active effort to help redefine the team roles, with team members having different functional and clinical backgrounds and responsibilities. Depending on the type of pathology, neurologists, oncologists, radiotherapists, radiologists, intervention radiologists, orthopedic surgeons, etc. can now be part of the team. The different team composition will require a conscious effort to be effective, and the requirements outlined above could serve as a guideline.

Indeed, medical specialists should not only respect and trust each other but also work toward shared goals in order to provide complex and interdependent care [14]. Of course, each individual treatment team is different. For example, healthcare teams with high levels of authority differentiation have clearly allocated leadership roles, which tend to be occupied by the most senior member of a team. It is important to recognize, however, that the prevalence of entrenched hierarchies and deep-rooted conflict among healthcare professionals can hinder decision-making and undermine high-quality care as well as innovation [15, 16].

The following items could help improve team collaboration and innovation:

- *Setting a mutual goal.* Teams can only be effective when its individual members have a common understanding of the objectives and are also committed to them. Balancing treatment outcomes and complications of various treatment innovations with the aim to improve quality of life can motivate team members to adapt innovation and create new ones.
- *Goal interdependence.* The goal is achieved only by the full team, not by its individual members. Goal interdependence leads team members to act in a way that creates mutual benefit. If the outcome of the various treatments and quality of life are shared goals, issues related to patient treatment, including feedback as a team, must be achieved collectively.
- *Support for innovation.* Teams can be more innovative when heads of department expect and approve of innovation, support members (even when their attempts to innovate are not successful), and reward new ideas and their implementation. Within healthcare however, innovation should be carefully monitored and implemented in order to maintain optimal patient safety.
- *Task orientation.* This is a shared concern for excellence that stems from the mutual goal. Teams with a mutual goal set high performance standards, monitor their performance, and provide each other feedback.
- *A cohesive team.* Researchers see cohesion as creating an environment that enables members to challenge each other and the *status quo*. For multidisciplinary teams consisting of professionals with strong professional identities, effort should be made to create commitment to the team.
- *Strong communication.* Strong internal communication (between team members) allows for sharing knowledge and ideas and creates a safe environment for providing feedback. External communication (communication with those outside the team) fosters innovation by learning from others and bringing new information into the team [17].

Conclusion

Innovation is vital to effective healthcare in highly demanding and competitive environments. Opportunities to innovate and to develop and implement skills in the workplace are central to the satisfaction of people at work. In neurosurgery, technical innovations are continuously introduced. Innovations from other disciplines will not only require good collaboration between healthcare specialists but also strong practical and cultural support to introduce new procedures and treatments. Establishing an open and safe working environment that encourages the development and implementation of innovation is a prerequisite.

References

1. Kohn LT, Corrigan JM, Donaldson MS. To err is human. Building a safer health care system. Washington D.C.: Institute of Medicine; 2000.
2. Andel C, Davidow SL, Hollander M, Moreno DA. The economics of health care quality and medical errors. J Health Care Finance. 2012;39(1):39–50.
3. Wagner C, Smits M, van Wagtendonk I, Zwaan L, Lubberding S, Merten H, Timmermans DRM. Oorzaken van incidenten en onbedoelde schade in ziekenhuizen. Een systematische analyse met PRISMA op afdelingen Spoedeisende Hulp (SEH), chirurgie en interne geneeskunde. [Causes of incidents and unwanted damage in hospitals. A systematic analysis with PRISMA on emergency, surgery and internal medicine departments]. EMGO Instituut en NIVEL; 2008.
4. Wong BM, Ackroyd-Stolarz S, Bukowskyj M, Calder L, Ginzburg A, Microys S, et al. The CanMEDS 2015 patient safety and quality improvement expert working group report. Ottawa; 2014.
5. Snyder J, Schultz L, Walbert T. The role of tumor board conferences in neuro-oncology: a nationwide provider survey. J Neuro-Oncol. 2017;133(1):1–7.
6. Gayed BA, Youssef R, Darwish O, Kapur P, Bagrodia A, Brugarolas J, et al. Multi-disciplinary surgical approach to the management of patients with renal cell carcinoma with venous tumor thrombus: 15 year experience and lessons learned. BMC Urol. 2016;16(1):43.
7. Cypko MA, Stoehr M, Kozniewski M, Druzdzel MJ, Dietz A, Berliner L, et al. Validation workflow for a clinical Bayesian network model in multidisciplinary decision making in head and neck oncology treatment. Int J Comput Assist Radiol Surg. 2017;12(11):1959–70.
8. Lamb BW, Sevdalis N, Benn J, Vincent C, Green JS. Multidisciplinary cancer team meeting structure and treatment decisions: a prospective correlational study. Ann Surg Oncol. 2013;20(3):715–22.
9. Johnson SB, Little TD, Masyn K, Mehta PD, Ghazarian SR. Multidisciplinary design and analytic approaches to advance prospective research on the multilevel determinants of child health. Ann Epidemiol. 2017;27(6):361–70.
10. Hewett DG, Watson BM, Gallois C, Ward M, Leggett BA. Intergroup communication between hospital doctors: implications for quality of patient care. Soc Sci Med. 2009;69(12):1732–40.
11. Manser T. Teamwork and patient safety in dynamic domains of healthcare: a review of the literature. Acta Anaesthesiol Scand. 2009;53(2):143–51.
12. Taplin SH, Weaver S, Salas E, Chollette V, Edwards HM, Bruinooge SS, et al. Reviewing cancer care team effectiveness. J Oncol Pract. 2015;11(3):239–46.
13. Madrid HP, Totterdell P, Niven K. Does leader-affective presence influence communication of creative ideas within work teams? Emotion. 2016;16(6):798–802.
14. Hollenbeck JR, Beersma B, Schouten ME. Beyond team types and taxonomies: a dimensional scaling conceptualization for team description. Acad Manag Rev. 2012;37(1):82–106.
15. Weller J, Boyd M, Cumin D. Teams, tribes and patient safety: overcoming barriers to effective teamwork in healthcare. Postgrad Med J. 2014;90(1061):149–54.
16. West MA, Lyubovnikova J. Illusion of team working in health care. J Health Organ Manag. 2013;27(1):134–42.
17. Hulsheger UR, Anderson N, Salgado JF. Team-level predictors of innovation at work: a comprehensive meta-analysis spanning three decades of research. J Appl Psychol. 2009;94(5):1128–45.

Culture and Attitudes Supporting Ethical Innovation in Neurosurgery

17

Marjel van Dam and Marike L. D. Broekman

Introduction

Advances in medicine and innovation are intertwined. Innovation is generally regarded as a novel practice that clearly departs from standard care. Whereas newly developed drugs are subjected to extensive rules and protocols, the body of regulations for surgical innovations is far less stringent. Many newly developed techniques are introduced outside the context of a randomized controlled trial and are not subject of strict oversight.

In 2007, the Institute of Medicine has proposed a learning health system (LHS). This has been described as a system in which "knowledge generation is so embedded into the practice of medicine that it is the natural product of the healthcare delivery process and leads to the continuous improvement of care" [1]. LHS has been suggested to aid the evaluation of innovation in (neuro)surgery. For example, the use of electronic health records (EHRs) would allow for assessment of innovative techniques and can help identify innovations that improve outcomes or possibly require a more systematic assessment [2].

M. van Dam (✉)
Intensive Care Center, University Medical Center Utrecht, Utrecht, The Netherlands

Center for Research and Development of Education, University Medical Center Utrecht, Utrecht, The Netherlands
e-mail: m.j.vandam@umcutrecht.nl

M. L. D. Broekman
Department of Neurosurgery, Computational Neurosciences Outcomes Center (CNOC), Brigham and Women's Hospital, Harvard Medical School, Boston, MA, USA

Department of Neurosurgery, Haaglanden Medical Center, The Hague, The Netherlands

Department of Neurosurgery, Leiden University Medical Center, Leiden, The Netherlands
e-mail: m.broekman@haaglandenmc.nl; m.l.d.broekman@lumc.nl

© Springer Nature Switzerland AG 2019
M. L. D. Broekman (ed.), *Ethics of Innovation in Neurosurgery*,
https://doi.org/10.1007/978-3-030-05502-8_17

LHS places a moral emphasis on learning, which requires a specific culture and attitudes. This chapter will discuss how a "learning" culture, which is characterized by trust, respect, collaboration, and transparency among others, can be created as well as the attitudes are needed for such a culture.

Creating a Culture Supporting Ethically Sound Innovation

Even though individual surgeons are often initiators of innovative treatments and inventors of novel devices that could benefit their practice, innovation is in general a joint venture [3]. In various settings, including business and medicine, it is evident that innovation can be enhanced through collaboration and teamwork.

Even though effective team collaboration depends on many factors, team leaders play a crucial role in creating and fostering an environment of innovation and initiative. Unfortunately, there is no clear road map that helps create such an environment. However, guidelines for team leaders in business management that aim to create an optimal environment for innovation might be useful for medical leaders as well. In 2014, Llopis formulated five suggestions for leaders to enable innovation (Table 17.1) [4], which will be discussed here.

The first piece of advice for team leaders is to trust yourself enough to trust others. Innovation requires willingness to leave behind old rules of thought and adopt new ones. This requires that each of the members of the team must become more transparent than before. Trust in oneself is essential to achieve this [4].

Besides trust in oneself, it is essential to trust the other members of the team. All team members should be confident that the other team members are available and accessible and that they pursue the same goals. This trust in oneself and in each other is essential for a culture in which innovation can thrive.

The second piece of advice is to collaborate and discover. Confidence in the team is essential, but without true collaboration within the team and with others, it will be impossible to develop something novel that can be implemented in the clinic. Collaboration not only means working together but also includes the identification of clear roles of all involved and pursuing the same goals. A true joint venture will make the odds for innovations higher.

Third, team leaders should stimulate communication. This is essential to build trust and to collaborate to innovate. By communicating and challenging each other's ideas in a respectful manner, innovations can be improved, and seeds can be planted for future innovations. The nature of the communication among team members is

Table 17.1 Six advices for team leaders to foster an environment of innovation and initiative (Adapted from [4])

1. Trust yourself enough to trust others
2. Collaborate and discover
3. Communicate to learn
4. Be a courageous change agent
5. Course correct to perfect
6. Create a culture of respect

critically important and should be constant, bidirectional, clear, and unambiguous. Only then can potential problems or misunderstandings be addressed immediately and innovation be supported.

Fourth, leaders that want to create an environment that promotes innovation have to be courageous agents of change. This means that leaders must constantly challenge everyone to think about how an innovation (be it novel procedures or devices for surgery) could be improved even further. Indeed, to develop effective strategies, teams should be built to learn from and incorporate both positive and negative experiences [5].

Last, to perfect the team and its innovative abilities, a leader needs to be able to course correct. This will steer a team more toward a culture that you aim to create and will keep the team members active, alert, and able to adjust to new circumstances, new people, and personalities. This makes building or developing a team, also known as "teaming," a dynamic process [6].

In our view, the creation of optimal circumstances to innovate requires one more crucial aspect: leaders need to foster a culture of respect [7]. A team whose members do not have mutual respect for each other will fail to move an idea forward effectively. A recent study showed that rudeness not only had an adverse effect on diagnostic and intervention parameters but also weakened collaborative processes such as information and workload sharing, helping, and communication – all of which are essential for patient care and safety but also for innovation [8].

At the same time, research in aviation has shown that positive attitudes about teamwork and respect for the work in the team are associated with error-reducing behavior [9]. Indeed, a culture of teamwork, transparency, respect, and communication can facilitate innovation and continuous improvement of patient outcomes.

Attitudes for Ethical Innovation

A culture that supports ethical innovation requires specific attitudes of the team members. Collaboration, interpersonal communication, leadership, and decision-making are indispensable skills for innovation and are increasingly recognized as important skills in surgery [10]. In order to incorporate these nonsurgical skills, certain attitudes and behaviors are essential and include non-macho behavior, transparency, and verifiable and sensitive behavior.

The Federal Aviation Administration and the Canadian Air Transport Administration identified "hazardous attitudes" (macho, impulsive, antiauthority, resignation, invulnerable, and confident) contributing to road traffic incidents among college-aged drivers and felt to be useful for the prevention of aviation accidents. A team of orthopedic surgeons has investigated whether these "hazardous attitudes" were prevalent among their colleagues [11]. Indeed, among the responding orthopedic surgeons, "hazardous attitudes" were common. In addition, almost one fifth of the responders implied absence of a safe climate. These attitudes do not contribute to an environment promoting ethical innovation.

In neurosurgery, an adapted version of the orthopedic surgeons' survey showed that high levels of "hazardous attitudes" are not as prevalent among neurosurgeons [12].

However, 12.2% of the responders had a potentially hazardous score for at least one of the evaluated attitudes. Potentially hazardous levels of resignation and anxiety were most prevalent. One could argue that certain attitudes are hazardous in an aviation or traffic setting but might very well be necessary for a neurosurgeon to function in a high-intensity environment. For instance, possessing a certain degree of machismo, antiauthority, and self-confidence may be required from a leader in times of emergency. These attitudes might offer stability and direction to a team in a stressful situation. However, these same attitudes might result in a situation in which members of the team don't dare to speak up or admit errors and are not eager to learn from them. Therefore, even in times of stress, attitudes should be guided by the rules of professional conduct or professionalism—the skill, good judgment, and polite behavior—that is expected from a person who is trained to do a job well. Even though it has been argued that too strict "rules of professionalism" might stifle innovation, professionalism does not have to interfere with innovative ideas or treatments. On the contrary, professionalism can give directions on how to support (ideas for) new treatments or techniques. In addition, it has been shown to help team performance: controlled supervision in a training situation resulted in significantly better performing teams than rude supervision [8].

Ultimately, professional attitudes will contribute to effective team collaboration, which is essential for innovation.

Conclusion

Innovation takes place without strict oversight and regulation, and ethical innovation requires a specific culture that focuses on team collaboration. Effective teamwork can be aided by medical leadership and requires specific attitudes guided by the rules of professionalism.

References

1. Olsen L, Aisner D, McGinnis JM. The learning healthcare system: workshop summary (IOM Roundtable on Evidence-Based Medicine). Washington DC: National Academies Press; 2007. p. xi–xvi.
2. Broekman ML, Carrière ME, Bredenoord AL. Surgical innovation: the ethical agenda: a systematic review. Medicine (Baltimore). 2016;95(25):e3790.
3. Wuchty S, Jones BF, Uzzi B. The increasing dominance of teams in production of knowledge. Science. 2007;316(5827):1036–9.
4. Llopis G. 5 ways leaders enable innovation in their teams, Forbes April 7 2014, [Internet]. [cited 2017 June 25]. https://www.forbes.com/sites/glennllopis/2014/04/07/5-ways-leaders-enable-innovation-in-their-teams/#1a7025208c4c.
5. Edmondson AC. Teaming: how organizations learn, innovate, and compete in the knowledge economy. San Francisco, CA: Jossey-Bass; 2012.

6. Nawaz H, Edmondson AC, Tzeng TH, Saleh JK, Bozic JK, Saleh KJ. Teaming: an approach to the growing complexities in health care. J Bone Joint Surg Am. 2014;96:e184(1–7.
7. Leape LL, Shore MF, Dienstag JL, et al. A culture of respect: II. Creating a culture of respect. Acad Med. 2012;87:853–8.
8. Riskin A, Erez A, Foulk TA, Riskin-Geuz KS, Ziv A, Sela R, Pessach-Gelblum L, Bamberger PA. Rudeness and medical team performance. Pediatrics. 2017;139(2):e20162305.
9. Helmreich RL, et al. Cockpit resource management: exploring the attitude–performance linkage. Aviat Space Environ Med. 1986;57:1198–200.
10. Yule S, Flin R, Paterson-Brown S, Maran N. Non-technical skills for surgeons in the operating room: a review of the literature. Surgery. 2006;139:140–9.
11. Bruinsma WE, Becker SJE, Guitton TG, Kadzielski J, Ring D. How prevalent are hazardous attitudes among orthopaedic surgeons? Clin Orthop Relat Res. 2015;473:1582–9.
12. Muskens I, van der Burgt SME, Senders JT, Lamba N, Peerdeman SM, Broekman ML. Behavior and attitudes among European neurosurgeons – An international survey. J Clin Neurosci. 2018;55:5–9.

Perspective: Future of Innovation in Neurosurgery

Marike L. D. Broekman

Ethical Innovation in Neurosurgery

In the past century, the practice of neurosurgery has changed dramatically. Not only the introduction of the microscope and bipolar forceps but also the invention of antibiotics, imaging techniques, and other developments that have changed how we practice medicine have made a profound impact on our profession. Nobody could have predicted the long-term effects of these discoveries at the time of invention. Most of these innovations were introduced without much oversight or regulation, and it would not be hard to guess that some important discoveries would not have made it to the clinic if there would have been strict oversight. Therefore, some have argued that innovation in neurosurgery should not be stifled by oversight and regulation.

However, innovations differ from standard care. Even though it remains a challenge to give a precise definition of innovation, it implies that the procedure or device is new and non-validated. This results in a situation of unknown (long-term) risks and benefits.

A potential solution to assess the (unknown) risks and benefits associated with a novel procedure would be to submit a research protocol. However, this would create an enormous amount of extra work, and the burden of bureaucracy could possibly grind innovation to a halt. Perhaps submitting a research protocol for every innovation does not have to be the best scenario as innovation is fundamentally different from

M. L. D. Broekman (✉)
Department of Neurosurgery, Computational Neurosciences Outcomes Center (CNOC), Brigham and Women's Hospital, Harvard Medical School, Boston, MA, USA

Department of Neurosurgery, Haaglanden Medical Center, The Hague, The Netherlands

Department of Neurosurgery, Leiden University Medical Center, Leiden, The Netherlands
e-mail: m.broekman@haaglandenmc.nl; m.l.d.broekman@lumc.nl

© Springer Nature Switzerland AG 2019
M. L. D. Broekman (ed.), *Ethics of Innovation in Neurosurgery*,
https://doi.org/10.1007/978-3-030-05502-8_18

research. Innovation does not aim to create generalizable knowledge but aims to benefit the individual patient instead. Surgeons indeed have the responsibility to act in the best interests of their patients. Therefore, one can assume that the intention of the surgeon trying a new procedure for their patient is good. However, despite such best intentions, the long-term risks and benefits are inherently unknown for non-validated procedures, which makes some form of evaluation or oversight essential.

However, as surgeons also have the responsible commitment to contribute to knowledge for future patients, it is of paramount importance to formulate ways to ethically introduce novel procedures and devices without hampering innovation.

Ethical introduction of novel surgical techniques might be helped using a structured approach, as, for instance, the IDEAL (Idea, Development, Exploration, Assessment, Long-term Follow-up) framework [1]. This framework has been developed by the IDEAL Collaboration—consisting of surgeons and methodologists—that strives to improve surgical research, with a special focus on innovation, and to overcome methodological challenges inherent to surgery [1]. Surgical innovations typically pass through five stages that are described by the framework. The framework describes the characteristics at each stage and proposes appropriate study designs [1]. Previously, we have shown that two recent innovations in neurosurgery (namely, the WEB device and endonasal endoscopic meningioma surgery) were not introduced according to this framework [2]. This might suggest that the IDEAL road map may not always be the perfect answer for ethically sound introduction of novel devices and procedures in neurosurgery, but the framework does hold great potential for the introduction and evaluation of at least some innovations in neurosurgery.

Unfortunately, it remains challenging to define which innovative procedures merit introduction according to the framework. Some have suggested that the introduction of novel procedures and devices (per a predefined road map) should depend on the level of potential risk to patients. Specifically, when the risks of an innovation can be considered low, oversight and adherence to a road map could be less strict [3].

Identification of trial-worthy innovations could be helped by using "big data." The use of electronic health records (EHRs) and digitized patient data increases the opportunities to use data that is collected as part of routine care for innovation and quality improvement purposes. This is in line with the concept of learning health system (LHS), which was introduced in 2007 by the Institute of Medicine. In this health system, knowledge generation is so embedded into the core of practice of medicine that it is a natural outgrowth and product of the healthcare delivery process and leads to continual improvement in care [4]. Key components of the LHS include a moral emphasis on learning, the search for alternatives to large randomized controlled trials, implementation of system databases, and fostering understanding of evidence-based medicine. In an LHS, there is no longer a clear distinction between care and research, and LHSs have the potential to translate research findings faster into clinical practice, thus empowering both doctors and patients with up-to-date knowledge.

The departure from the clear distinction between research and care requires a unique ethics framework. To this end, Faden et al. proposed an ethics framework that identifies seven obligations for all parties involved in an LHS [5]. These include (1) to respect the rights and dignity of patients, (2) to respect the clinical judgment of clinicians, (3) to provide optimal care to each patient, (4) to avoid imposing non-clinical risks and burdens on patients, (5) to reduce health inequalities among populations, (6) to conduct responsible activities that foster learning from clinical care and clinical information, and (7) to contribute to the common purpose of improving the quality and value of clinical care and healthcare systems. Altogether, these obligations "constitute a necessary condition, within a learning health care system, of an adequate ethics. In the absence of any one of these obligations, the framework would lose a basic norm, rendering the framework deficient" [5].

The way these obligations have been formulated in this ethics framework puts a significant moral emphasis on learning and the consequent requirement of a specific culture and novel professional attitudes, including specifically, transparency, willingness to learn, and open communication. When properly executed, this framework has the promise to have a positive impact on innovation and implementation of novel procedures and to help minimize the potential risks associated with blurring the traditional boundaries between research and care.

Conclusion

Ethical innovation in neurosurgery can be stimulated not only by predefined road maps for the ethical introduction of novel procedures and techniques such as the IDEAL framework but also by the introduction of so-called learning health systems. In such LHSs, the traditional boundaries between research and care may be blurred, requiring a specific ethics framework, with obligations for all parties involved as well as certain professional attitudes. These will help minimize potential risks associated with blurring the traditional boundaries between research and care and improve ethical innovation in neurosurgery.

References

1. McCulloch P, Altman DG, Campbell WB, et al. No surgical innovation without evaluation: the IDEAL recommendations. Lancet. 2009;374(9695):1105–12. https://doi.org/10.1016/S0140-6736(09)61116-8.
2. Muskens IS, Diederen SJH, Senders JT, et al. Innovation in neurosurgery: less than IDEAL? A systematic review. Acta Neurochir. 2017;159(10):1957–66. https://doi.org/10.1007/s00701-017-3280-3.
3. Largent EA, Joffe S, Miller FG. Can research and care be ethically integrated? Hast Cent Rep. 2011;41(4):37–46.
4. Olsen L, Aisner D, McGinnis JM. The learning healthcare system: workshop summary (IOM Roundtable on Evidence-Based Medicine). Washington, DC: National Academies Press; 2007. p. xi–xvi.
5. Faden RR, Kass NE, Goodman SN, Pronovost P, Tunis S, Beauchamp TL. An ethics framework for a learning health care system: a departure from traditional research ethics and clinical ethics. Hast Cent Rep. 2013;Spec No:S16–27.

Multidisciplinary Team Meetings

Neurosurgeons are trained to treat complex (and rare) pathologies for which often no guidelines or standardized options are available. This has resulted in the increased use of multidisciplinary team meetings (MDMs) involving professionals from diverse backgrounds to focus on specific patient cases, discuss appropriate treatment options, and provide optimal care for the individual patient. An effective MDM requires time (to ensure all relevant questions can be addressed), resources, and structured communication including giving adequate feedback on the resulting outcome.

MDMs are generally valued for their contribution to clinical decision-making, education, and improved access to clinical trials [5]. Recently, a multidisciplinary team approach has been shown to lower complication rates, for example, in patients with renal cell carcinoma with venous tumor thrombus [6]. There is a growing recognition of the importance of decision-making in multidisciplinary teams [7], and certain interventions (e.g., optimizing communication structure, introduction of checklists, team training, and providing written guidance) can improve decision-making and ultimately improve patient care [8].

Some of the benefits of MDMs are the sharing of up-to-date knowledge about new treatment options. Even so, there is still a need for clear, evidence-based clinical practice guidelines for the conduct of MDMs, with accepted standards and objective measures of performance [9].

Teamwork

Medical specialists tend to have a strong sense of professional identity, mostly defined by their specialization [10]. Conflicts between different specialists can arise, which can have a profoundly negative influence on the quality and safety of care. Therefore, adequate (i.e., multidisciplinary) teamwork is of great importance. Several models of good teamwork have been described in the literature, highlighting typical characteristics of what constitutes an effective team [11, 12]. In order to ensure effective communication and interaction between the members of the MDM team, it is worth highlighting the following key attributes:

- *Cohesion of collaboration.* Mutual respect, developing trust, and the desire to work together in the future; individuals should feel their work is essential to the team, their roles should be meaningful, and their contributions should be identifiable.
- *Shared mental models.* Strength of shared goals, shared perception of a situation, and shared understanding of team structure, team task, and team roles.
- *Coordination.* Adaptive coordination and identification of triggers indicating that key steps have been completed or are in progress.
- *Cooperation.* Demonstrating uncompromising commitment to working collaboratively.